Springer Theses

Recognizing Outstanding Ph.D. Research

Aims and Scope

The series "Springer Theses" brings together a selection of the very best Ph.D. theses from around the world and across the physical sciences. Nominated and endorsed by two recognized specialists, each published volume has been selected for its scientific excellence and the high impact of its contents for the pertinent field of research. For greater accessibility to non-specialists, the published versions include an extended introduction, as well as a foreword by the student's supervisor explaining the special relevance of the work for the field. As a whole, the series will provide a valuable resource both for newcomers to the research fields described, and for other scientists seeking detailed background information on special questions. Finally, it provides an accredited documentation of the valuable contributions made by today's younger generation of scientists.

Theses are accepted into the series by invited nomination only and must fulfill all of the following criteria

- They must be written in good English.
- The topic should fall within the confines of Chemistry, Physics, Earth Sciences, Engineering and related interdisciplinary fields such as Materials, Nanoscience, Chemical Engineering, Complex Systems and Biophysics.
- The work reported in the thesis must represent a significant scientific advance.
- If the thesis includes previously published material, permission to reproduce this must be gained from the respective copyright holder.
- They must have been examined and passed during the 12 months prior to nomination.
- Each thesis should include a foreword by the supervisor outlining the significance of its content.
- The theses should have a clearly defined structure including an introduction accessible to scientists not expert in that particular field.

More information about this series at http://www.springer.com/series/8790

Monika Johanna Kreitmair

The Effect of Uncertainty on Tidal Stream Energy Resource Estimates

Doctoral Thesis accepted by
University of Edinburgh, United Kingdom

Springer

Author
Dr. Monika Johanna Kreitmair
Department of Engineering
University of Cambridge
Cambridge, UK

Supervisors
Prof. Ton S. van den Bremer
University of Oxford
Oxford, UK

Prof. Alistair G. L. Borthwick
The University of Edinburgh
Edinburgh, UK

ISSN 2190-5053 ISSN 2190-5061 (electronic)
Springer Theses
ISBN 978-3-030-57657-8 ISBN 978-3-030-57658-5 (eBook)
https://doi.org/10.1007/978-3-030-57658-5

This Springer imprint is published by the registered company Springer Nature Switzerland AG
The registered company address is: Gewerbestrasse 11, 6330 Cham, Switzerland

The actual science of logic is conversant at present only with things either certain, impossible, or entirely doubtful, none of which (fortunately) we have to reason on. Therefore the true logic for this world is the calculus of Probabilities, which takes account of the magnitude of the probability which is, or ought to be, in a reasonable man's mind.

James Clerk Maxwell [1850]

Dedicated to Li Li. Thank you for the kindness you showed us in sharing with us your city and your home. You are missed.

Supervisors' Foreword

Uncertainty may be defined as 'a state of having limited knowledge where it is impossible to exactly describe an existing state or future outcome' (Douglas Hubbard, *How to measure anything: Finding the value of intangibles in business*, John Wiley & Sons, 2007). Uncertainty affects almost all areas of environmental fluid mechanics, including weather systems, hydrological analysis, flood flow prediction and risk assessment, sediment transport and geomorphological change, water quality, and wind and hydropower assessment.

In hydro-kinetic power assessment, uncertainties in bed roughness and turbine drag characteristics can greatly affect power estimates. This, in turn, can have appreciable impact on optimization of arrays of hydro-kinetic turbines which depends on the local hydrodynamic conditions, basin dynamics, array configuration, blockage, etc. The maximum power extracted by a turbine fence or array is not related directly to the local undisturbed kinetic flux or the natural energy dissipation rate due to bed friction, but includes a feedback effect from the energy extracted by the turbines. Accounting for uncertainty within this feedback is crucial.

At river or coastal basin-scale, numerical models for estimating power yield from a site have been developed based on the depth-integrated shallow water equations. Within these, flows are modelled as long waves and the effect of turbines usually represented either as enhanced bed roughness or by linear momentum actuator disk theory, or by means of an electrical circuit analogy. Although important advances have occurred in our theoretical understanding of the power available to hydro-kinetic turbines in shallow flows and considerable advances are taking place in shallow flow simulation, power predictions are subject to considerable uncertainty arising from observation and parameter errors, parameter uncertainty, and model assumptions. For example, estimates of power available from the Pentland Firth between the north mainland of Scotland and the Orkneys have varied from about 2 to 14 GW depending on the modelling approach used and the choice of input parameters.

This thesis uses the mathematics of stochastic processes and numerical modelling techniques to show that significant uncertainty can arise from insufficient field data for model calibration and the natural variability of the flow environment.

Three methods are used to propagate uncertainty: transfer of a probability density function (pdf) in bed friction coefficient through a closed-form analytic solution for power (useful for benchmark purposes); an equivalent discrete pdf transfer method through numerically determined solutions; and an approximate, perturbation method that provides insight into the leading-order effect of input uncertainty. The methodologies are tested for unidirectional flow in a straight channel containing a fence of turbines, where it is found that the expected power is sensitive to the channel dynamics, turbine blockage, and site geometry. It is also predicted that expected power increases with bed friction uncertainty in a channel fully spanned by turbines, but reduces in laterally unconfined conditions. The optimal number of turbines is lower when placed in a fully-spanned channel and higher when the flow is laterally unconfined, for bed friction uncertainty. By examining uncertainty in turbine drag, it is shown that the expected power increases when turbines are tuned to a higher drag than predicted by a deterministic model. The discrete pdf transfer method is then applied to tidal stream power assessment for the Pentland Firth, using information supplied by an open-source finite-element shallow flow solver. The approach is particularly useful in cases where the exact functional dependence of power on an input random variable of interest is unknown and the power response surface can be determined from numerical model simulations. The discrete pdf transfer method is essentially a discrete analogy to the 'derived distribution approach' that has seen extensive application in engineering hydrology. In the case of the Pentland Firth, the approach represents a step towards proper characterization of the available tidal stream power resource. The methods for uncertainty transfer presented in the thesis could readily be applied to many other problems encountered in hydraulic engineering, such as flow routing, urban flood risk, and reservoir sedimentation. In the words of Ilya Prigogine (1917–2003), a physical chemist who was awarded the Nobel Prize for Chemistry in 1977, "The future is uncertain… but this uncertainty is at the very heart of human creativity".

Oxford, UK
Edinburgh, UK
September 2020

Ton S. van den Bremer
Alistair G. L. Borthwick

Abstract

In river and tidal stream power assessment, uncertainties arise from model assumptions and the inexact specification of physical and numerical model parameters. Combined, such uncertainties can greatly affect power estimates for a given site. The thesis examines the effects of bed roughness and turbine drag uncertainties on turbine power estimates.

An analytic model is developed for transfer of bed friction uncertainty to power extracted from turbines in a strait, representative of a river. A validated finite volume solver of the shallow water equations is developed and applied to simulate flow driven by a constant head difference through a one-dimensional strait. The presence of a turbine fence is included using enhanced bed friction. A parameter study examines the effect of uncertainty propagation from bed friction to power. Excellent agreement is obtained between the analytic and numerical power uncertainty estimates for a given input bed friction PDF.

Perturbation methods are used to determine the leading-order effect of bottom friction uncertainty in tidal stream power assessment. The theoretical models consider quasi-steady flow in a channel completely spanned by tidal turbines, a similar channel but retaining the inertial term, and a circular turbine farm in laterally unconfined flow. It is found that changes to expected power depend on the dynamic balance in the channel, the turbine configuration, and the geometry of the site considered. Bottom friction uncertainty increases estimates of expected power in a fully-spanned channel, but has the reverse effect in laterally unconfined farms. The optimal number of turbines under bottom friction uncertainty is lower for a fully-spanned channel and higher in laterally unconfined farms. The effect of uncertainty in turbine drag is also considered.

A standard methodology is presented for uncertainty propagation using general computational models. The methodology is tested using a shallow flow model of the Pentland Firth, where power statistics are determined according to input bed friction probability distribution, and the results compared against those from the (simplified) analytic perturbation approaches. Although the analytic models for channels perform reasonably well regarding the estimate of expected power, the predictions from the unconfined analytic model were not so satisfactory owing to the model assumptions.

The methods for uncertainty transfer presented in the thesis could readily be applied to many other problems encountered in hydraulic engineering, such as river flow routing, urban flood risk, reservoir sedimentation, etc.

Acknowledgements

First and foremost, my thanks goes to my supervisors Prof. Alistair G. L. Borthwick and Prof. Ton S. van den Bremer. Their insight and kindness have left a deep impression on me and I hope to one day live up to the great examples they set. I am immensely grateful to Prof. Scott Draper (University of Western Australia) and Prof. Tom Adcock (University of Oxford) for discussions which helped to refine the analysis in several aspects of my thesis and for the opportunity to work with them. I would also like to thank Prof. Vladimir Nikora and Dr. Thanasis Angeloudis for the (dare I say) enjoyable examination at the end. Funding for this research was supported by the Engineering and Physical Sciences Research Council (EPSRC), grant no. R44708.

In the time it took to research, analyse, and write up the contents of this thesis, I have had the fortune to meet many amazing people without whom I would not have made it through the Ph.D. or, at least, with a great deal less sanity. I want to thank Joe for his continuing (albeit possibly unwilling) support, as well as Paul B., Nacho, Kaswar, Paul, Encarni, Anna, Daniel F., my office mates and the numerous other people who enriched my time in Edinburgh. I am grateful to Grayling and Alice who managed to inveigle their way into my life through the cunning use of roast dinners. Thanks goes to the Edinburgh University Muay Thai club for a continual source of catharsis. Tea and Simon especially saw to it that I would never go too long without either physical harm or cheesecake. Thank you to the frenemies from my MSc cohort, Dave, Dan, Ricardo, and Abel, who still manage to irritate me beyond reason today. Thank you to all of the wonderful people who made our visit to Beijing so special, particularly Lin Jie (Tsinghua University) and Li Li (Peking University) who went far beyond the call of duty to make us feel welcome. Thank you to Guy and (the unquestionably Parisienne but nominally Australian) Anas for the time we spent together in Perth and beyond. In Oxford, I want to thank Chris for the time he deigns to spend with me and Anela from whom I draw a great deal of inspiration, despite the short period of time for which I have known her. Finally, thank you to Charlotte, Sarah, Jack, and Paul: I am truly grateful for having you in my life, in spite of the infrequency of my communication.

Lastly (and by no means least-ly) I would like to thank my family. My parents who have supported me throughout my academic meanderings, and my sisters who, through experiences in their own Ph.D.s and with benefit of hindsight, were always able to tell me what I needed to hear.

Contents

Nomenclature

A	Channel cross-sectional area
a	Amplitude of sinusoidal tide
A_{turb}	Turbine frontal area
C	Chézy friction coefficient
C_d	Quadratic bed roughness coefficient
C_t	Turbine drag coefficient
E[x]	Expectation operator applied to x
f	Coriolis parameter due to Earth's rotation
$f_x(x)$	Probability density function for random variable x
F_{turb}	drag force due to presence of turbine
g	Acceleration due to gravity
h	Total water depth
λ_0	Non-dimensional bed roughness parameter and measure of the dynamic balance between natural friction and inertial forces acting in a channel
λ_T	Non-dimensional equivalent drag force produced by the presence of turbines in a channel
L	Channel length
μ	Dynamic viscosity
μ_x	Arithmetic mean of random variable x
ν_T	Depth-scaled turbulent eddy viscosity
ν_t	Turbulent eddy viscosity
n	Manning friction coefficient
ω	Angular frequency of sinusoidal tide
Ω_E	Angular frequency of rotation of the Earth
ϕ	Angle of latitude
p	Pressure
p_a	Atmospheric pressure at free surface $z = \zeta$
$\overline{P}_{\text{max}}$	Maximum average power over a tidal cycle
P_{KE}	Kinetic energy flux
Q	Volume flux along channel, $Q = Au$

Q_0	Undisturbed volumetric flow rate
ρ	Fluid density
$\hat{\sigma}_x$	Relative standard deviation of random variable x, $\hat{\sigma}_x = \sigma_x/\mu_x$
σ_x	Standard deviation of random variable x
τ_b	Bed stress
$\mathbf{u}$	Depth averaged velocity vector, $\mathbf{u} = (u, v, w)^T$
$\mathbf{u}_*$	Instantaneous flow velocity vector, $\mathbf{u}_* = (u_*, v_*, w_*)^T$
u_0, u_e	Flow velocity at channel entrance and exit, respectively
W	Channel width
x	Cartesian coordinate system
ζ	Free surface elevation over still water level
z_b	Location of sea bed
z_0, Z_0	Roughness length. Capital refers to random variable, lowercase to realisation of random variable
$\hat{z}_0, \hat{Z}_0$	Ratio of roughness length to flow depth. Capital refers to random variable, lowercase to realisation of random variable

Chapter 1
Introduction and Literature Review

"We demand rigidly defined areas of doubt and uncertainty!"

— Douglas Adams,
The Hitchhiker's Guide to the Galaxy

Tides are extremely long ocean waves driven by the net force produced by gravitational attraction between sea water, the Earth, Moon, Sun, and other celestial bodies. As gravitational tides sweep across the surface of the Earth, the associated currents are altered by the effects of the Coriolis force, bed friction, and bathymetric features [52]. Acceleration of currents occurs at constrictions, such as in narrowing channels and around headlands [15]. In these localised, accelerated flows, the energy flux is often sufficiently large to allow economically viable extraction of energy from tidal currents using hydro-electric turbines.

1.1 Tidal Stream Power Assessment

Tidal power has a long history, particularly along the western seaboard of Europe where many tidal mills (essentially water wheels driven by a tidal head difference between ocean and mill pond) were installed from Roman times onward [12]. Perhaps the first reference in the literature to tidal stream power generation by turbines is by Sir Robert Kane in 1844 [32], who wrote of the tides at the coasts of Ireland that

> *"...the motion of the tide becomes a source of power, and tide mills form an important variety of water mills. In England they are scarcely used, coals being so cheap, but to us, by proper application, I am convinced they may become an important basis of industry."*

M. J. Kreitmair, *The Effect of Uncertainty on Tidal Stream Energy Resource Estimates*, Springer Theses, https://doi.org/10.1007/978-3-030-57658-5_1

and further

> *"The turbine is, therefore, peculiarly the machine for economizing tidal power. For each acre of reservoir it may be expected to give at least three horse-power working twenty hours out of the twenty-four."*

In more recent years, concerns about the impact of increasing greenhouse gas emissions from fossil fuels have driven research into technologies for the exploitation of alternative, less carbon-intensive energy sources. Tidal energy is of particular interest because it is, at least theoretically, predictable [3] and, despite being intermittent and localised, offers a reliable energy source. The mechanisms of the origin of the tides are very well established. Tidal prediction algorithms were developed as early as the late 19th century, which informed mechanical devices that calculated tidal elevations using 10 or more harmonic constituents of the tidal forcing.

Tidal energy may be split into two categories depending on origin of energy extracted: tidal stream, where power is extracted from the motion of tidally driven flow, and tidal range, where power is generated from the head difference produced by tidal elevation changes. Over the last decade, there has been rapid development of techniques aimed towards assessment of available tidal stream resource, both in industry and academia. Early attempts to assess tidal stream power were based on the undisturbed kinetic flux through the area swept by a turbine farm, $P_{\mathrm{KE}} = 1/2 A_{\mathrm{turb}} \rho u^3$, where A_{turb} is the frontal area of the turbine farm, ρ is the density of water and u is the undisturbed flow velocity [26].]P_{KE}Kinetic energy flux A_{turb}Turbine frontal area Examples include surveys by Black & Veatch Consulting Ltd [9] for the Pentland Firth, United Kingdom, [61] for the Minas Passage, Canada, and [25] for the eastern coasts of North America. The kinetic flux approach for estimating the power potential at basin scale has been shown to be flawed because the power in tidal waves is not given simply by the kinetic energy flux across a plane [40]. This approach neglects the contribution from the static energy in the system resulting from the potential head difference. Furthermore, energy is dissipated as heat due to turbulent interactions which is also not accounted for. Garrett and Cummins [19] showed that there is no simple relationship between the undisturbed kinetic flux in a tide-driven channel and the maximum power that may be extracted from the flow because the undisturbed kinetic flux does not account for the feedback effect of energy extraction on flow velocities [21]. At very low levels of turbine deployment (i.e. drag on the flow), very little power is extracted. This power increases as further turbines are added and drag on flow is increased, provided the feedback effect on the flow remains negligible. On the other hand, at very high levels of drag on the flow due to the presence of a large number of turbines, the flow speed is reduced to such an extent that almost no power is extracted because the flow has become 'choked'. Hence, there exists an optimum turbine deployment that maximises the power extracted by the turbines through balancing the flow deceleration and drag deployed. Three approaches have been suggested to account for the presence of tidal turbines in shallow water basin models: enhanced roughness; linear momentum actuator disc theory; and an electrical analogy. Each of these is examined in turn below.

The first approach considers enhanced bed roughness to represent the effect of the tidal turbines whereby the power dissipated by the additional resistance equates to the power removed by the turbines. Garrett and Cummins [19] found that the maximum average power over a tidal cycle in a channel connecting two infinitely large basins (such that energy extraction does not affect water levels in the basins) is given by

$$\overline{P}_{\max} = \gamma \rho g a Q_0, \tag{1.1}$$

where γ is a non-dimensional multiplier whose value depends on the dynamic balance of the flow in the channel, from $\gamma = 0.21$ for inertia and $\gamma = 0.24$ for friction-dominated flows, ρ is water density, g is acceleration due to gravity, a is the amplitude of the head difference either end of the channel, and Q_0 is the amplitude of the undisturbed (i.e. without turbines present) flow rate. Here, the presence of the turbines is modelled as a patch of enhanced friction uniformly distributed throughout the channel, as the turbines are assumed to fully span the width of the channel. Consequently, this approach is not able to take into account energy losses due to wake mixing, i.e. turbulent energy mixing losses due to pressure gradients as the by-pass flow around the turbines recombines with the slower wake flow.

The second approach involves representing turbines as porous discs and using linear momentum actuator disc theory (LMADT) to determine the head loss across the turbines and hence the power extracted. LMADT originated from analysis of thrust on ship propellors in the late 19th century and was later applied to determine power extracted by wind turbines in the early 20th century [8, 31, 35]. More recently it has been applied for tidal stream turbines in the early 21st century. For wind, water stream, and tidal flow turbines, principles of mass, momentum, and energy conservation are selectively applied across control volumes, with turbines introduced as line sinks of momentum in the flow. The head difference across the momentum sink is evaluated, and energy extracted determined. In the context of tidal stream turbines, [20] calculated the power from turbines in a channel with low Froude number, necessitated by considering a volume-constrained channel subject to the rigid-lid approximation. This model was extended by Vennell [62, 63] by combining it with the one-dimensional model of Garrett and Cummins [19] to determine optimal turbine tuning (flow velocity through a turbine) and farm efficiency. By using a method of scale separation to decompose the flow structures around individual turbines and the fence of turbines as a whole, [47, 48] extended the volume-constrained model to determine power extracted by a fence only partially spanning a channel. Similarly, [66, 67] applied actuator disc models to model power from an infinite row of turbines in an open channel which allowed for free surface deformation. Houlsby et al. [29] and Houlsby and Vogel [30] extended this theory to include wake mixing. Using this approach, the available power (i.e. for electricity generation) rather than extractable power (power removed from the channel), can be distinguished.

A third approach for determining power from a channel, which allows for sub-channels, is by using an analogy with a electrical circuit, [44, 51, 53]. In this analogy, the multiply connected channel network is replaced by an equivalent circuit where driving head is represented by an alternating voltage (i.e. a potential difference), flow rate by an alternating current, water mass by inductance, and energy dissipation due to natural bed roughness or the energy extraction by turbines as a resistance. Using this approach [16] produced an equivalent circuit for the Pentland Firth, which provided quick assessments of tidal power for different device configurations.

These approaches have been applied to numerous sites throughout the world to assess numerically the power available from tidal flows and the impact on the existing flow regime. Examples include assessment for the Pentland Firth, U.K. [2, 16, 17, 43, 49], the Anglesey Skerries, U.K. [56], Johnstone Strait, Canada [59], the Bay of Fundy, Canada [33], the Malacca Strait, Malaysia [11], the Alas Strait, Indonesia [10], the Shannon Estuary, Ireland [18, 45], and São Marco Bay, Brazil [24].

The forgone discussion has focused chiefly on the modelling of tidal stream energy extraction at basin scale. However, a major challenge of tidal stream resource assessment is that hydrodynamic interactions take place across a large range of scales, from the blade scale (on the order of a few centimeters), over device scale (tens of metres), to basin, and finally regional and far-field scale (hundreds of kilometres) [3]. Blade scale modelling requires high fidelity computational fluid dynamics (CFD) approaches in order to model correctly boundary layers on the blades, lift and drag forces, and, consequently, structural loading and fatigue. At the turbine scale, methods range from actuator disk approaches (where the turbine is represented by a single disk of momentum sink, e.g. [29]), blade element momentum (BEM) methods (where the rotor disk is split into annular rings to account for the change in velocity from blade root to tip, e.g. [50, 65]), and actuator line approaches (blades are resolved individually as lines of momentum sink with particular angular velocities, e.g. [4, 13]).

While no fully commercial tidal stream power farms exists at the time of writing, important steps have been taken in developing the nascent tidal stream industry. Testing and demonstration sites, such as the grid-connected European Marine Energy Centre (EMEC) site in the Orkney Isles or the FORCE site in the Bay of Fundy, exist to test and demonstrate in situ the readiness of turbines and determine behavioural characteristics. The first large-scale tidal energy site has been deployed as part of the MeyGen tidal energy project, the first phase of which consists of four 1.5 MW turbines installed in the Pentland Firth (completed in 2018), delivering power to the mainland grid [57].

As noted above, an alternative method to tidal stream energy extraction to harness energy from tides is by exploiting tidal range through using tidal barrages. These utilise the head difference between high and low tide to drive turbines situated within barriers separating a body of water from the remainder of the ocean. By strategically closing and opening gates and allowing water to flow past turbines, tidal barrages produce energy at either flood, or ebb, or both flood and ebb in the tidal cycle, thereby providing highly predictable power. An example is the La Rance tidal power station in France, which has a rated capacity of 240 MW. However, high capital investment costs and significant ecological impacts [5] have meant that proposals for tidal range projects are often turned down, as has been the case for the Severn Barrage and Swansea lagoon in the UK. While outwith the scope of the present thesis, the study of the environmental impacts and power potential of tidal barrages is a field of ongoing research (see for example [46]).

1.2 Uncertainty in Environmental Flows

Uncertainty (1350–1400 Middle English *uncerteynte*) involves a state of limited knowledge or inherent randomness where it is impossible exactly to describe the present state and future outcome(s). The following poem encapsulates uncertainty and its gradations as perhaps understood in medieval times:

One who knows and knows that he knows…
His horse of wisdom will reach the skies
One who knows, but doesn't know that he knows…
He is fast asleep, so you should wake him up!
One who doesn't know, but knows that he doesn't know…
His limping mule will eventually get him home.
One who doesn't know and doesn't know that he doesn't know…
He will be eternally lost in his hopeless oblivion!

Four types of men, Ibn Yamin
Persia, 13th century

The sentiment of the poem is echoed 700 years later in a 2002 speech by the then U.S. Secretary of Defense, Donald Rumsfeld:

> *"...as we know, there are known knowns; there are things we know we know. We also know there are known unknowns; that is to say we know there are some things we do not know. But there are also unknown unknowns–the ones we don't know we don't know. And if one looks throughout the history of our country and other free countries, it is the latter category that tend to be the difficult ones."*

Evidently, uncertainty is ubiquitous both in the political and the physical realms. While it can never be entirely eliminated, it is possible to ascend from 'unknown unknowns' to 'known unknowns' by carefully cataloguing and quantifying the uncertainty associated with our predictions of the world around us.

In the context of modelling, uncertainty is often sorted into one of two categories [34]. The first is aleatory uncertainty, and it arises from the intrinsic randomness of the physical world. The word 'aleatory' stems from the Latin *alea*, which means games of chance or rolling of dice—one may remember the phrase attributed to Julius Caesar upon his crossing of the Rubicon river: *alea iacta est*: the die is cast. The second category is epistemic uncertainty, which originates from a lack of knowledge or data concerning the physical system being considered. The word comes from the Greek $\epsilon\pi\iota\sigma\tau\eta\mu\eta$ (episteme), meaning 'good knowledge' (but can also refer to 'science' and 'understanding'). The sources of uncertainty outlined in the next subsection for numerical simulations, fall predominantly into the realm of epistemic uncertainty and can, as such, be reduced as information becomes available and measuring techniques improve.

1.2.1 Sources of Uncertainty

Numerical simulations of a physical system will inevitably give predictions that differ from observations. The discrepancy arises from fundamental limitations in a model's ability to describe perfectly all characteristics of the system that is being modelled. These limitations are manifested as errors and uncertainties and may be grouped into three main types, following [36]:

I - Model Error In expressing the physics of a system in a mathematical form, it is often necessary to apply simplifying assumptions to make a problem tractable. The assumption result in loss of information such that the model will not perfectly reproduce the behaviour exhibited in the real world, leading to errors and uncertainty in the model predictions. In the context of tidal flow modelling, an example of this occurs in the derivation of the shallow water equations where a reduction in dimensionality of the system takes place due to integration over the depth of the flow, such that the complexity of the three-dimensional coordinate system is projected onto a more readily solvable (as well as computationally less intensive) two-dimensional coordinate system.

II - Numerical error In order to solve a mathematical model numerically, governing equations are discretised and resolved on a computational mesh. This approximation of continuously-valued functions at a finite number of discrete points introduces 'discretisation' or 'truncation' error. Reduction of this error may be achieved by computational mesh refinement (both spatial and temporal), albeit at the expense of increased computing time. Moreover, due to the finite memory space of the computer used for the numerical solution, the number of digits to which data may be represented is finite, and so any exact mathematical value, which theoretically has infinitely many digits, is rounded. Designs of effective numerical methods specifically aim to reduce such round-off errors, for example through ensuring convergence of the mathematical model.

III - Data error To make predictions site specific, the mathematical model requires input data such as system geometry (in the case of oceanographic modelling, bathymetric data and geometry of coastal features), boundary and initial conditions (flow velocities, water depth, etc.) and other physical parameters or model constants (bed roughness, gravitational acceleration, etc.). To obtain these, field measurements need to be collected for the system of interest. In a marine environment, acquisition of such measurements is notoriously difficult, particularly within the context of tidal stream energy conversion because the very attribute that makes a site a candidate for development, namely fast flow speed, creates rough, turbulent flow which impedes characterisation of necessary input data. Compounding this problem is the inherent uncertainty of the system: nonlinear interactions of tidally driven flow with domain boundaries can produce turbulent flow structures and chaotic responses.

1.2.2 Uncertainty in Tidal Models

At coastal-scale, most tide hydrodynamic models are based on the nonlinear shallow water equations. Specific to these models, uncertainties arise from several sources, including the inexact specification of the natural environment (due to lack of accurate field data on the tidal velocity field, turbulence, large-scale eddying motions, etc.), the physical model parameters (bed roughness, bathymetry, boundary and initial conditions, etc.), model assumptions, and numerical parameters (grid resolution, time step, etc.). Combined, these uncertainties can give rise to considerable discrepancy between different power estimates for a given site. For example, predictions of the average power available from the Pentland Firth, UK, one of the most promising sites for tidal stream energy extraction in the world, span more than an order of magnitude (from 0.62 GW [1] to 9 GW [41]), with little consensus as to the true power potential [17]. Table 1.1 provides a summary of power values quoted in the literature. It should be noted that the estimates quoted were made using different methodologies and survey scopes, which act to contribute significant levels of uncertainty (see *Model error*, in Sect. 1.2.1 above). The variation apparent in the power estimates is due to the fact that tidal stream energy is still in the early stages of development, and standardised methods do not yet exist for establishing the power potential of a site.

Of the sources of uncertainty listed above, the bed friction coefficient C_d is particularly important. C_dQuadratic bed roughness coefficient In practice, this parameter is often used to tune numerical models based on the shallow water equations, so that they predict water levels and velocity vectors in agreement with observations at relatively sparse spatial locations. Various researchers have carried out sensitivity analyses for different values of drag coefficient applied uniformly throughout the domain. For example, in a power resource assessment of the Pentland Firth, [2] examined the sensitivity of tidal stream power estimates to the value of bed friction coefficient C_d in the range [0.0025, 0.01], applied uniformly through the flow

Table 1.1 Estimates of the maximum power potential of the Pentland Firth. 'Average power' indicates power averaged over a tidal cycle

Study	Average power (GW)	C_d	Method/comments
[9]	1	–	Undisturbed flux
[40]	9	–	
[2]	1.9	0.005	LMADT approach. Spring-neap average
[16]	3.4	–	Electrical analogy
[17]	4.2	0.005	Enhanced bed roughness approach
[49]	5.3	–	3-D model with momentum sink
	Peak power (GW)		
[1]	0.62	–	"Exploitable" power
[55]	18	–	Fraction of dissipation by natural roughness
[64]	1.9	–	LMADT, M_2 tide forcing

domain. Adcock et al. [2] found that the available power reduced as C_d increased. However, the average power determined from the numerical model over the range of values of C_d considered was greater than the power calculated using the average value of C_d. That is, the dependence of power on C_d is non-linear. In addition, [2] found that no single value of C_d applied throughout the modelled domain produced results which matched the field measurements of both tidal phase and current magnitude for the Pentland Firth, and settled on a value of $C_d = 0.005$ in a compromise. In a similar study, [23] varied C_d from 0.002 to 0.010 and found a constant value of $C_d = 0.004$ (again applied uniformly throughout the domain) gave best agreement, while acknowledging the significant spatial heterogeneity of the seabed. Furthermore, the appropriate value of C_d depends also on the extent of the modelled domain and the location of the boundaries, with deeper flows being modelled the further out the boundary is taken. To fully acknowledge the spatially varying nature of the sea bed and account for effects due to domain extent, a spatially varying bed roughness coefficient would be needed.

An alternative view is that the bed friction coefficient should be hydraulically correct in terms of the boundary layer dynamics and not treated simply as a tuning parameter (see Salter [54]). Soulsby [58] lists a range of values ($C_d \in [0.0011, 0.0043]$ for silt-sand to rippled sand) that could be applied to different marine bed surfaces, and which deal with skin friction and form drag. In short, there is a lack of agreement as to which bed friction values should be applied. Culley et al. [14] performed a sensitivity analysis of estimated power from an optimised tidal farm in the Inner Sound of the Pentland Firth, which highlighted the significant influence of the value of the bottom friction coefficient on the numerical results. In the study, estimated values for power reduced as the bed roughness increased near the farm, and, at sufficiently high values of local bed friction, the flow began to bypass the farm along paths of lower frictional resistance.

Given the various sources of uncertainty, it is important to ascertain how these individually and collectively impact on tidal model predictions, as highlighted by the discrepancy exhibited in tidal power estimates in Table 1.1, for example. Consequently, predictions should become more reliable, and the uncertainty itself could play a useful role in optimising tidal turbine layouts.

1.2.3 Modelling Uncertainty Propagation

The ubiquity of uncertainty in the natural world and the need to design reliably and robustly has generated a large body of work in the field of uncertainty propagation covering a wide range of applications [37]. Methods for investigation of the impact that uncertain input parameters have on the model output can be broadly classed into two categories: non-intrusive and intrusive methods. This classification depends on how the method extracts the stochastic nature of the system from the deterministic governing equations.

I - Non-intrusive methods These methods, also known as 'simulation-based' or 'sampling' methods, retain the governing equations in deterministic form that then function as a 'black-box' through which various samplings of the stochastic parameter field are run [36]. An archetypal example of this is provided by Monte Carlo (MC) simulations. Conceptually straightforward, the MC approach generates a large number of realisations of the random parameter field. Each realisation is then solved deterministically (i.e. given as input to the black-box) and the results then post-analysed to derive the statistical distribution of the model solution. The advantages of this approach are its robustness and the ease of its application to both linear, nonlinear and even chaotic problems. However, in order to resolve adequately high frequency spatio-temporal fluctuations in random parameters, MC simulations often require fine numerical grids making the computational cost of each realisation very high. Moreover, to ensure sufficient convergence of the output moments to their theoretical ensemble values, a large number of realisations needs to be performed (typically depending on the degree of heterogeneity of the system). This leads to a significant computational burden that can be prohibitive. Refinements to the MC approach have been developed to reduce this problem of computational expense by applying the method to a finite subset of realisations [39]. This subset needs to be chosen in such a way as to ensure that the resulting predictions are representative of the global behaviour of the model solution space.

One such method is 'adaptive sampling' where the sampling choice of subsequent computer runs is iteratively guided by analysis, using statistical prediction models, of the runs performed with the existing samples [42]. Another method, known as 'importance sampling', has been applied to flow and transport in random porous media by Lu and Zhang [39]. Therein, values of the random input variables that have greater impact on the model solution (as determined by importance density functions) are sampled at greater frequency, thus requiring fewer samples overall. The simulation outputs are then analysed to eliminate bias introduced by the weighted sampling. With appropriately chosen importance density functions, this technique can be several orders of magnitude more efficient than conventional MC simulations.

II - Intrusive methods Also known as 'analytic' methods, intrusive methods alter the nature of the governing equations to reflect the stochastic nature of the system they describe through various decompositions of the stochastic parts of the input variables and terms in the equations. These decompositions may then be used to derive the statistical nature of the solution directly, instead of having to analyse an ensemble of simulated output. The advantage of such techniques over simulation-based ones is their computational efficiency. However, these too suffer from the so-called"curse of dimensionality", coined by Bellman [7], in that the computational complexity grows geometrically with increasing numbers of stochastic parameters. These methods may be classed into several subcategories:

Local expansion-based methods, which include Taylor series, perturbation methods, etc. These methods are advantageous when dealing with relatively small input variability and outputs that do not express high nonlinearity but perform poorly against the large variability of inputs and nonlinearity of performance functions [37]. A method of this type was utilised by Horritt [27] to investigate the impact of topographic uncertainty in one-dimensional shallow water flow by applying a Taylor expansion to the governing equations and truncating after the second-order term. The results were in good agreement to those obtained from Monte Carlo simulations, and it was found that higher-order terms remained negligible provided the bed perturbations were small compared to the flow depth. It was noted, however, that the sufficiency of second-order expansion was not confirmed for a two-dimensional model.

Functional expansion-based methods have enjoyed growing attention in recent years. In particular the polynomial chaos expansion has become widely used for uncertainty representations, stochastic mechanics, solution of stochastic differential equations, etc. [37]. First developed by Wiener [68] for Gaussian processes and generalised by Xiu and Karniadakis [69] to the most common statistical distributions, polynomial chaos expansion (PCE) expresses the stochastic variables in the governing differential equations as a combination of orthogonal polynomials and reduces the problem of propagating uncertainties to the task of determining the expansion coefficients.

PCE has been applied to the shallow water equations and compared favourably against MC simulations for simple scenarios of long-wave propagation over a hump in one dimension [22]. This offers a promising step towards application of this method to more complicated scenarios. In oceanography, PCE has been used by Thacker et al. [60] to explore the propagation of uncertainties in the boundary conditions in the flow from the Caribbean Sea into the Gulf of Mexico and how these manifest themselves as uncertainties in the surface elevation field. Due to absence of comprehensive data on both the open boundary conditions and their uncertainties, [60] reduced the spatially and temporally varying multivariate description of the inflow to a small number of random variables. Thacker et al. [60] concluded that a better approach to dealing with lack of data would be to determine boundary conditions and associated uncertainty from a large ensemble of runs of an outer model, defining the input at the boundary. This, however, would be prohibitive in terms of computational cost.

Finally, it should be noted that the calculation of solution statistics through this technique can be significantly compromised by the presence of discontinuities in random variable dimensions [6].

1.2.4 Concluding Remarks

There exist many sources of uncertainty in tidal stream power assessment, of which a very important and relatively under-researched source is bed roughness. Uncertainty propagation in shallow flow models is an area of active research, though primarily through local-expansion-based and MC methods applied to flood risk rather than tidal stream power assessment, e.g. [22, 28]. To date, tidal stream resource assessment has received no formal treatment using uncertainty propagation methods and there is clearly scope for applying methods for uncertainty propagation within tidal stream assessment. As the marine renewable energy sector develops, it may be of interest to look to the wind energy sector for an established procedure for determining uncertainty predictions on energy yield [38].

1.3 Aim and Objectives

Aim

This thesis aims to quantify uncertainty propagation through the shallow water equations with turbines represented through enhanced bed friction, using analytical and numerical techniques, and demonstrate application of the approach to an idealised channel and the Pentland Firth.

Objectives

- To identify and quantify sources of uncertainty in tidal stream power assessment.
- To develop analytical models for assessing the effect of bed roughness uncertainty on hydropower estimates for steady and tidal oscillatory flow in a one-dimensional channel, and tidal oscillatory flow in a two-dimensional basin.
- To develop a validated Godunov-type finite volume solver of the one-dimensional shallow water equations, and carry out a parameter study of the effect of uncertainty in bed roughness on hydro-kinetic power in a one-dimensional channel (for comparison against the analytical model).
- To use an industry standard open source continuous Galerkin solver, ADCIRC, to examine bed friction uncertainty on estimates of power extraction from the Pentland Firth, Scotland.

1.4 Synopsis

This chapter discusses the present state of knowledge regarding stream power assessment in coastal waters, summarises methods available for uncertainty propagation, and lists the main aim and objectives of the thesis. Chapter 2 presents a brief derivation of the shallow water equations, which underpin the work undertaken throughout the thesis, and are commonly used by engineers to predict free surface flows in shallow lakes, rivers, and coastal basins. Chapter 3 outlines the derivation for power from a fence of hydro-kinetic devices in steady flow (as in a river) and describes a Godunov-type finite volume solver of the one-dimensional shallow flow equations which is verified against standard benchmark tests. This is subsequently used in a parameter study of the effect of uncertainty in bed friction on the expected power and its variance for hydro-kinetic turbines. Chapter 4 provides derivations of analytical formulae for assessment of uncertain power for tidal stream turbines considering uncertainty in natural bed friction. Chapter 5 extends the concept to consider the effect of uncertain drag on power. Chapter 6 presents results from a numerical modelling study of the Pentland Firth, where numerical predictions of power statistics are compared against estimates from the simplified analytical tools derived in Chap. 4. Chapter 7 lists the main findings of the thesis, and makes recommendations for future research.

Publications Arising from This Thesis

- Kreitmair M. J., Borthwick A. G. L., and van den Bremer T. S. (2017). Effect of bed roughness uncertainty on tidal stream power estimates. *Proceedings of the 12th European Wave and Tidal Energy Conference (EWTEC)*, Cork, Ireland.
- Kreitmair M. J., Borthwick A. G. L., van den Bremer T. S., and Draper S. (2018). The effects of uncertain bottom friction on estimates of tidal current power. *Proceedings of the* 6th *Oxford Tidal Energy Workshop*, Oxford, United Kingdom.
- Kreitmair M. J., Draper S., Borthwick A. G. L., and van den Bremer T. S. (2019). The effect of uncertain bottom friction on estimates of tidal current power. *Royal Society Open Science*, **6**(1):180941.
- Kreitmair M. J., Adcock T. A. A., Borthwick A. G. L., Draper S., and van den Bremer T. S. *(In preparation)*. The effect of bed roughness uncertainty on tidal stream power estimates for the Pentland Firth.
- Kreitmair M. J., Borthwick A. G. L., and van den Bremer T. S. (2019). Uncertainty Quantification for Tidal Power in the Pentland Firth. *Proceedings of the* 15th *Young Coastal Scientists and Engineers Conference*, Glasgow, United Kingdom.

References

1. ABPmer (2007) Quantification of exploitable tidal energy resources in UK waters. Technical Report R.1349, ABP Marine Environmental Research, commissioned by npower Juice
2. Adcock TAA, Draper S, Houlsby GT, Borthwick AGL, Serhadlioğlu S (2013) The available power from tidal stream turbines in the Pentland Firth. Proc Royal Soc A 469:1471–2946
3. Adcock TAA, Draper S, Nishino T (2015) Tidal power generation - A review of hydrodynamic modelling. Proc Inst Mech Eng A 1–17
4. Apsley DD, Stallard T, Stansby PK (2018) Actuator-line CFD modelling of tidal-stream turbines in arrays. J Ocean Eng Mar Energy 4(4):259–271
5. Baker C (1991) Tidal power. Energy Policy 19(8):792–797
6. Barth TJ, Griebel M, Keyes DE, Nieminen RM, Roose D, Schlick T (2013) In: Bijl H, Lucor D, Mishra S, Schwab C (eds) Uncertainty quantification in computational fluid dynamics, vol 92. Lecture notes in computational science and engineering. Springer International Publishing, Cham
7. Bellman R (1961) Adaptive control processes. Princeton University Press, Princeton, NJ, United States of America
8. Betz A (1920) Das Maximum der theoretisch möglichen Ausnützung des Windes durch Windmotoren. Zeitschrift für das gesamte Turbinenwesen 26:307–309
9. Black & Veatch Consulting Ltd (2005) Phase II - UK tidal stream energy resource assessment. Technical Report 0, report prepared for the Carbon Trust Marine Energy Challenge
10. Blunden LS, Bahaj AS, Aziz NS (2013) Tidal current power for Indonesia? An initial resource estimation for the Alas Strait. Renew Energy 49:137–142
11. Bonar PAJ, Schnabl Andrea M, Lee W-K, Adcock TAA (2018) Assessment of the Malaysian tidal stream energy resource using an upper bound approach. J Ocean Eng Mar Energy 4:99–109

12. Cartwright DE (2000) Tides: a scientific history. Cambridge University Press. Cambridge
13. Creech AC, Borthwick AG, Ingram D (2017) Effects of support structures in an LES actuator line model of a tidal turbine with contra-rotating rotors. Energies 10(5)
14. Culley DM, Funke SW, Kramer SC, Piggott MD (2015) Tidal stream resource assessment through optimisation of array design with quantification of uncertainty. In: Proceedings of the 11th European wave and tidal energy conference (EWTEC)
15. Draper S (2011) Tidal stream energy extraction in coastal basins. PhD thesis, University of Oxford
16. Draper S, Adcock TAA, Borthwick AGL, Houlsby GT (2013) An electrical analogy for the Pentland Firth tidal stream power resource. Proc R Soc A: Math Phys Eng Sci, 470
17. Draper S, Adcock TAA, Borthwick AGL, Houlsby GT (2014) Estimate of the tidal stream power resource of the Pentland Firth. Renew Energy 63:650–657
18. Fallon D, Hartnett M, Olbert A, Nash S (2014) The effects of array configuration on the hydro-environmental impacts of tidal turbines. Renew Energy 64:10–25
19. Garrett C, Cummins PF (2005) The power potential of tidal currents in channels. Proc Royal Soc A 461:2563–2572
20. Garrett C, Cummins PF (2007) The efficiency of a turbine in a tidal channel. J Fluid Mech 588:243–251
21. Garrett C, Cummins PF (2008) Limits to tidal current power. Renew Energy 33(11):2485–2490
22. Ge L, Cheung K, Kobayashi M (2008) Stochastic solution for uncertainty propagation in nonlinear shallow-water equations. J Hydraul Eng 134(December):1732–1743
23. Gillibrand P, Walters R, McIlvenny J (2016) Numerical simulations of the effects of a tidal turbine array on near-bed velocity and local bed shear stress. Energies 9(10):852
24. González-Gorbeña E, Rosman PC, Qassim RY (2015) Assessment of the tidal current energy resource in São Marcos Bay, Brazil. J Ocean Eng Mar Energy 1:421–433
25. Hagerman G, Polagye B (2006) Methodology for estimating tidal current energy resources and power production by tidal in-stream energy conversion (TISEC) devices. Technical Report EPRI - TP - 001 NA Rev 3, Electric Power Research Institute
26. Hammons TJ (1993) Tidal power. Proc IEEE 8(3):419–433
27. Horritt M (2002) Stochastic modelling of 1-D shallow water flows over uncertain topography. J Comput Phys 180(1):327–338
28. Horritt M (2006) A linearized approach to flow resistance uncertainty in a 2-D finite volume model of flood flow. J Hydrol 316(1–4):13–27
29. Houlsby GT, Draper S, Oldfield M (2008) Application of linear momentum actuator disc theory to open channel flow. Technical report, University of Oxford
30. Houlsby GT, Vogel CR (2017) The power available to tidal turbines in an open channel flow. Proc Inst Civil Eng - Energy 170:12–21
31. Joukowsky N (1920) Windmill of the NEJ type. In: Transactions of the Central Institute for Aero-hydrodynamics of Moscow, 1(57)
32. Kane R (1844) The industrial resources of Ireland
33. Karsten RH, Swan A, Culina J (2013) Assessment of arrays of in-stream tidal turbines in the Bay of Fundy. Philos Trans R Soc A, Math Phys Eng Sci, p 371
34. Kiureghian AD, Ditlevsen O (2009) Aleatory or epistemic? Does it matter? Struct Saf 31(2):105–112
35. Lanchester FW (1915) A contribution to the theory of propulsion and the screw propeller. Nav Eng J 27(2):509–510
36. Le Maître OP, Knio OM (2010) Spectral methods for uncertainty quantification. Scientific computation. Springer, Netherlands, Dordrecht
37. Lee SH, Chen W (2009) A comparative study of uncertainty propagation methods for black-box-type problems. Struct Multidiscip Optim 37(3):239–253
38. Livermore S (2015) Wave and tidal energy yield uncertainty - refernce document. Technical Report PN000083-SRT-005, Catapult Offshore Renewable Energy
39. Lu Z, Zhang D (2003) On importance sampling Monte Carlo approach to uncertainty analysis for flow and transport in porous media. Adv Water Resour 26(11):1177–1188

40. MacKay DJC (2007) Under-estimation of the UK tidal resource. http://www.inference.org.uk/sustainable/book/tex/TideEstimate.pdf
41. MacKay DJC (2008) Sustainable energy - without the hot air. UIT Cambridge
42. Maljovec D, Wang B, Kupresanin A, Johannesson G, Pascucci V, Bremer P-T (2013) Adaptive sampling with topological scores. Int J Uncertain Quantif 3(2):119–141
43. Martin-Short R, Hill J, Kramer SC, Avdis A, Allison Pa, Piggott MD (2015) Tidal resource extraction in the Pentland Firth, UK: potential impacts on flow regime and sediment transport in the Inner Sound of Stroma. Renew Energy 76:596–607
44. Miles JW (1971) Resonant response of harbours: an equivalent-circuit analysis. J Fluid Mech 46(02):241–265
45. Nash S, O'Brien N, Olbert A, Hartnett M (2014) Modelling the far field hydro-environmental impacts of tidal farms - a focus on tidal regime, inter-tidal zones and flushing. Comput Geosci 71:20–27
46. Neill SP, Angeloudis A, Robins PE, Walkington I, Ward SL, Masters I, Lewis MJ, Piano M, Avdis A, Piggott MD, Aggidis G, Evans P, Adcock TA, Židonis A, Ahmadian R, Falconer R (2018) Tidal range energy resource and optimization - past perspectives and future challenges. Renew Energy 127:763–778
47. Nishino T, Willden RHJ (2012) The efficiency of an array of tidal turbines partially blocking a wide channel. J Fluid Mech 708:596–606
48. Nishino T, Willden RHJ (2013) Two-scale dynamics of flow past a partial cross-stream array of tidal turbines. J Fluid Mech 730:220–244
49. O'Hara Murray R, Gallego A (2017) A modelling study of the tidal stream resource of the Pentland Firth, Scotland. Renew Energy 102:326–340
50. Olczak a, Stallard T, Feng T, Stansby PK (2016) Comparison of a RANS blade element model for tidal turbine arrays with laboratory scale measurements of wake velocity and rotor thrust. J Fluids Struct 64:87–106
51. Prandle D (1980) Modelling of tidal barrier schemes: an analysis of the open-boundary problem by reference to AC circuit theory. Estuar Coast Mar Sci 53–71
52. Pugh DT (1996) Tides, surges and mean sea-level (reprinted with corrections). Wiley, Chichester, United Kingdom
53. Rainey RCT (2009) The optimum position for a tidal power barrage in the Severn estuary. J Fluid Mech 636:497
54. Salter SH (2009) Correcting the Under-estimate of the Tidal-Stream Resource of the Pentland Firth. In:Proceedings of the 8th European wave and tidal energy conference (EWTEC)
55. Salter SH, Taylor JRM (2007) Vertical-axis tidal-current generators and the Pentland Firth. Proc Inst Mech Eng Part A - J Power Enerrgy 22:181–99
56. Serhadlioğlu S, Adcock TAA, Houlsby GT, Draper S, Borthwick AGL (2013) Tidal stream energy resource assessment of the Anglesey Skerries. Int J Mar Energy 3–4:98–111
57. SIMEC Atlantis Energy (2018) Tidal projects: Meygen. https://simecatlantis.com/projects/meygen/. Accessed: 2018-11-29
58. Soulsby RL (1997) Dynamics of marine sands. Thomas Telford Publications, London, United Kingdom
59. Sutherland G, Foreman M, Garrett C (2007) Tidal current energy assessment for Johnstone Strait, Vancouver Island. Proc Inst Mech Eng A 221(2):147–157
60. Thacker W, Srinivasan a, Iskandarani M, Knio O, Hénaff ML (2012) Propagating boundary uncertainties using polynomial expansions. Ocean Model 43–44:52–63
61. Triton Consultants Ltd (2006) Canada ocean energy atlas (phase 1) potential tidal current energy resources analysis background. Technical report, prepared for the Canadian Hydraulics Centre, Natural Resources Canada
62. Vennell R (2010) Tuning turbines in a tidal channel. J Fluid Mech 663:253–267
63. Vennell R (2011a) Estimating the power potential of tidal currents and the impact of power extraction on flow speeds. Renew Energy 36(12):3558–3565
64. Vennell R (2011b) Tuning tidal turbines in-concert to maximise farm efficiency. J Fluid Mech 671:587–604

65. Vogel CR, Willden RH, Houlsby GT (2018) Blade element momentum theory for a tidal turbine. Ocean Eng 169(April):215–226
66. Whelan J, Thomson M, Graham JMR, Peiró J (2007) Modelling of free surface proximity and wave induced velocities around a horizontal axis tidal stream turbine. In: Proceedings of the 7th European wave and tidal energy conference, Porto, Portugal
67. Whelan JI, Graham JMR, Pieró J (2009) Inertia effects on horizontal axis tidal-stream turbines. In: Proceedings of the 8th European wave and tidal energy conference (EWTEC), pp 586–591
68. Wiener N (1938) The homogeneous chaos. Am J Math 60(4):897–936
69. Xiu D, Karniadakis GE (2003) A new stochastic approach to transient heat conduction modeling with uncertainty. Int J Heat Mass Transf 46(24):4681–4693

Chapter 2
Derivation of the Shallow Water Equations

2.1 Introduction

This chapter presents a derivation of the shallow water equations, which describe flows where horizontal length-scales are much larger than the vertical. This occurs when hydrodynamic processes are small in comparison to gravity effects, and is therefore appropriate for tidal flows in channels where the tidal wavelength λ_w is greater than the channel depth h, such that $\lambda_w / h \ll 2\pi$.

The shallow water equations are commonly used in tidal hydrodynamic modelling [2] because they provide a greatly simplified set of governing equations, thereby improving ease of computation. This however comes at the cost of accuracy as to the resolution of vertical flow features. Depth-averaged models are unable to simulate secondary flows and shear stresses in the water column as they do not model vertical momentum transfer and thus do not account for phenomena resulting from these such as separation and horizontal mixing of flow due to complex bathymetries or the presence of tidal turbines [4, 10, 12]. However, in the absence of extreme stratification, bulk flow phenomena are well modelled and hence the shallow water equations provide a good estimate for tidal power resource assessments of large turbine arrays, for example, spanning the entire width of a channel because the power is dependent on bulk flow rather than local spatial gradients [2].

2.2 Shallow Water Equations

The shallow water equations may be derived by depth integrating the continuity and Navier-Stokes equations (see for example [5]) or from mass and momentum conservation analysis applied to a control volume of infinitesimal plan area extending through the whole depth of the fluid column (see Abbott [1]). The following is a brief presentation of the derivation following the former approach.

M. J. Kreitmair, *The Effect of Uncertainty on Tidal Stream Energy Resource Estimates*, Springer Theses, https://doi.org/10.1007/978-3-030-57658-5_2

For an incompressible and Newtonian fluid, the continuity equation $\partial\rho/\partial t + \nabla \cdot (\rho\mathbf{u}_*) = 0$ may be reduced to

$$\frac{\partial u_*}{\partial x} + \frac{\partial v_*}{\partial y} + \frac{\partial w_*}{\partial z} = 0, \tag{2.1}$$

where $\mathbf{u}_* = (u_*, v_*, w_*)^T$ is the three-dimensional instantaneous velocity vector with components in the $\mathbf{x} = (x, y, z)^T$ direction, and ρ is fluid density. This equation describes the conservation of mass in a fluid of constant and uniform density.

Conservation of momentum is commonly expressed by the Navier-Stokes equations. In non-conservative form, these may be expressed as

$$\frac{\partial u_*}{\partial t} + u_*\frac{\partial u_*}{\partial x} + v_*\frac{\partial u_*}{\partial y} + w_*\frac{\partial u_*}{\partial z} = X - \frac{1}{\rho}\frac{\partial p}{\partial x} + \frac{\mu}{\rho}\nabla^2 u_*, \tag{2.2a}$$

$$\frac{\partial v_*}{\partial t} + u_*\frac{\partial v_*}{\partial x} + v_*\frac{\partial v_*}{\partial y} + w_*\frac{\partial v_*}{\partial z} = Y - \frac{1}{\rho}\frac{\partial p}{\partial y} + \frac{\mu}{\rho}\nabla^2 v_*, \tag{2.2b}$$

$$\frac{\partial w_*}{\partial t} + u_*\frac{\partial w_*}{\partial x} + v_*\frac{\partial w_*}{\partial y} + w_*\frac{\partial w_*}{\partial z} = Z - \frac{1}{\rho}\frac{\partial p}{\partial z} + \frac{\mu}{\rho}\nabla^2 w_*, \tag{2.2c}$$

where t is time, p is pressure, μ the dynamic viscosity, and $\mathbf{X} = (X, Y, Z)^T$ represents the body force vector. The equations express the balance between local and convective acceleration of a fluid parcel, body forces, pressure gradient, and viscous momentum diffusion. For examples of a detailed derivation of the Navier-Stokes equations, see Hughes and Brighton [6] and White [13].

By multiplying the incompressible continuity equation (2.1) with u_* and adding the resulting expression to the x-momentum equation (2.2a), the conservative form of the momentum balance in the x-direction can be obtained after some rearranging:

$$\frac{\partial u_*}{\partial t} + \frac{\partial u_*^2}{\partial x} + \frac{\partial u_* v_*}{\partial y} + \frac{\partial u_* w_*}{\partial z} = X - \frac{1}{\rho}\frac{\partial p}{\partial x} + \frac{\mu}{\rho}\nabla^2 u_*. \tag{2.3}$$

Equivalent expressions may be obtained in the y- and z-directions.

The non-conservative momentum equations (2.2) are given in terms of primitive variables (velocity components) which are not conserved across discontinuities and do not hold in scenarios with discontinuous features. In contrast, while mathematically equivalent, the conservative x-momentum equation (2.3) (and equivalent expressions in the y- and z-directions) strictly adhere to the conservation law of momentum and hence hold across discontinuities such as hydraulic jumps. This makes the conservative formulation useful in numerical schemes.

2.2.1 Reynolds-Averaged Navier-Stokes Equations (RANS)

Because the flow in coastal basins is often turbulent (see Taylor [11]), the dependent flow variables, u_*, v_*, w_*, and p, may be separated into a time-averaged component and a randomly fluctuating component. This permits the equations to be manipulated into a form that is more tractable, following [9]. For example, the velocity component in the x-direction u, may be written as $u_* = \overline{u_*} + u'_*$, where the time-averaged component $\overline{u_*} = \frac{1}{\Delta t}\int_t^{t+\Delta t} u_*\, dt$, and the randomly fluctuating component is u'_*, with $\overline{u'_*} = \frac{1}{\Delta t}\int_t^{t+\Delta t} u'_*\, dt = 0$. The time period Δt is long compared to the timescale of the fluctuations but short compared to the tidal period. Inserting this decomposition into the continuity equation (2.1) and time-averaging gives

$$\frac{\partial \overline{u_*}}{\partial x} + \frac{\partial \overline{v_*}}{\partial y} + \frac{\partial \overline{w_*}}{\partial z} = 0. \tag{2.4}$$

Applying the same procedure to the conservative x-momentum equation (2.3), noting that $\overline{\overline{u_*} + u'_*} = \overline{u_*}$ and $\overline{(\overline{u_*} + u'_*)(\overline{u_*} + u'_*)} = \overline{u_*}\,\overline{u_*} + \overline{u'_* u'_*}$, gives, after some rearranging, the Reynolds-averaged Navier-Stokes (RANS) equation for incompressible turbulent flow in the x-direction:

$$\frac{\partial \overline{u_*}}{\partial t} + \frac{\partial \overline{u_*}^2}{\partial x} + \frac{\partial \overline{u_*}\,\overline{v_*}}{\partial y} + \frac{\partial \overline{u_*}\,\overline{w_*}}{\partial z} = \overline{X} - \frac{1}{\rho}\frac{\partial \overline{p}}{\partial x} + \frac{1}{\rho}\left[\frac{\partial \sigma'_{xx}}{\partial x} + \frac{\partial \tau_{xy}}{\partial y} + \frac{\partial \tau_{xz}}{\partial z}\right], \tag{2.5}$$

where

$$\sigma'_{xx} = \mu\frac{\partial \overline{u_*}}{\partial x} - \rho\overline{u'_* u'_*}, \quad \tau_{xy} = \mu\frac{\partial \overline{u_*}}{\partial y} - \rho\overline{u'_* v'_*}, \text{ and } \tau_{xz} = \mu\frac{\partial \overline{u_*}}{\partial z} - \rho\overline{u'_* w'_*}. \tag{2.6}$$

Here we observe Reynolds stress terms (second terms in (2.6)) that are quadratic in the fluctuating velocity components which represent interactions between the fluctuating turbulent quantities, in addition to the viscous stress terms (first terms in (2.6)). In general, the dispersive momentum transport due to turbulent eddies is much larger than due to laminar diffusion, such that the molecular viscosity contributions may be neglected.

The time-averaged body force components per unit mass, $\overline{X}$, $\overline{Y}$, and $\overline{Z}$ are given by combination of the Coriolis force acting in the xy-plane, and the effect of gravity in the downwards z-direction, i.e. $\overline{\mathbf{X}} = \overline{\mathbf{f}} \wedge \overline{\mathbf{u}_*} + \overline{\mathbf{g}}$ or (longhand)

$$\begin{pmatrix}\overline{X}\\ \overline{Y}\\ \overline{Z}\end{pmatrix} = \begin{pmatrix}0\\0\\f\end{pmatrix} \wedge \begin{pmatrix}\overline{u_*}\\ \overline{v_*}\\ \overline{w_*}\end{pmatrix} + \begin{pmatrix}0\\0\\-g\end{pmatrix} = \begin{pmatrix}f\overline{v_*}\\ -f\overline{u_*}\\ -g\end{pmatrix}, \tag{2.7}$$

where $\wedge$ is the cross product and $f = 2\Omega_E \sin\phi$ is the Coriolis parameter resulting from the Earth's rotation, with Ω_E being the angular frequency of rotation of the Earth and ϕ the angle of latitude.

The overbar notation is subsequently dropped to avoid clutter.

2.2.2 *Depth-Integration*

For shallow basins where the length scales are predominantly horizontal, and particularly for tidal flows where the vertical velocity component w_* is much smaller than its horizontal counterparts u_* and v_*, the Reynolds-averaged continuity and Navier-Stokes momentum equations may be integrated over the flow depth to give a simpler system of equations whose solution yields the depth-averaged velocity field and local water level in time and space.

Taking the bed to be at $z = -z_b$ and the free surface at $z = \zeta$, where the still water level is at $z = 0$ (see Fig. 2.1), integration of the Reynolds-averaged continuity equation (2.4) over the flow depth and application of Leibniz's rule gives

$$\frac{\partial}{\partial x}\int_{-z_b}^{\zeta} u_*\,\mathrm{d}z + \frac{\partial}{\partial y}\int_{-z_b}^{\zeta} v_*\,\mathrm{d}z - \left(u_*|_{\zeta}\frac{\partial \zeta}{\partial x} + u_*|_{-z_b}\frac{\partial(-z_b)}{\partial x}\right) - \left(v_*|_{\zeta}\frac{\partial \zeta}{\partial y} + v_*|_{-z_b}\frac{\partial(-z_b)}{\partial y}\right) + \left(w_*|_{\zeta} - w_*|_{-z_b}\right) = 0. \quad (2.8)$$

The following boundary conditions are applied. At $z = -z_b$ the bed is assumed solid, and a no-flow kinematic boundary condition applies i.e.

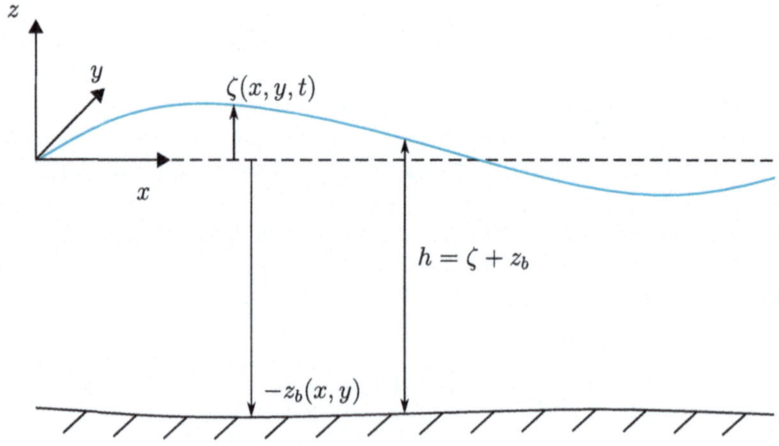

Fig. 2.1 Definition sketch for the shallow water equations

$$w_*|_{-z_b} = \frac{\mathrm{d}(-z_b)}{\mathrm{d}t} = \frac{\partial(-z_b)}{\partial t} + u_*|_{-z_b}\frac{\partial(-z_b)}{\partial x} + v_*|_{-z_b}\frac{\partial(-z_b)}{\partial y}. \tag{2.9}$$

That is, there is no flow through the solid bed. Similarly, at $z = \zeta$, the kinematic free surface boundary condition states that particles at the free surface remain there. Hence,

$$w|_\zeta = \frac{\mathrm{d}(\zeta)}{\mathrm{d}t} = \frac{\partial\zeta}{\partial t} + u_*|_\zeta\frac{\partial\zeta}{\partial x} + v|_\zeta\frac{\partial\zeta}{\partial y}. \tag{2.10}$$

Substituting the above conditions into the depth-averaged continuity equation (2.8) then gives

$$\frac{\partial(h)}{\partial t} + \frac{\partial}{\partial x}\int_{-z_b}^{\zeta} u_*\,\mathrm{d}z + \frac{\partial}{\partial y}\int_{-z_b}^{\zeta} v_*\,\mathrm{d}z = 0, \tag{2.11}$$

where $h = \zeta + z_b$. Defining the depth-averaged velocity components using the first mean value theorem as

$$u = \frac{1}{h}\int_{-z_b}^{\zeta} u_*\,\mathrm{d}z \quad \text{and} \quad v = \frac{1}{h}\int_{-z_b}^{\zeta} v_*\,\mathrm{d}z, \tag{2.12}$$

the depth-averaged continuity equation becomes

$$\frac{\partial h}{\partial t} + \frac{\partial(uh)}{\partial x} + \frac{\partial(vh)}{\partial y} = 0. \tag{2.13}$$

For a non-erodible bed, the position of the bed is constant with respect to time, $\partial(-z_b)/\partial t = 0$ such that $\partial h/\partial t = \partial\zeta/\partial t$ and (2.13) further reduces to

$$\frac{\partial\zeta}{\partial t} + \frac{\partial(uh)}{\partial x} + \frac{\partial(vh)}{\partial y} = 0. \tag{2.14}$$

Next consider the pressure acting on the fluid column. Neglecting vertical acceleration and shear stress terms, the z-component of the momentum conservation equation reduces to a balance between gravitational acceleration and the pressure gradient:

$$\frac{1}{\rho}\frac{\partial p}{\partial z} + g = 0. \tag{2.15}$$

Integration over the fluid depth gives the hydrostatic pressure distribution

$$p(z) = p_a + \rho g(\zeta - z), \tag{2.16}$$

where p_a is the atmospheric pressure at the free surface $z = \zeta$. Neglecting atmospheric pressure gradients, the spatial derivatives of pressure may then be expressed in terms of water elevations, i.e.

$$\frac{\partial p}{\partial x} = \rho g \frac{\partial \zeta}{\partial x}. \tag{2.17}$$

Repeating the depth-averaging process for the Reynolds-averaged Navier-Stokes x-momentum equation (2.5), and substituting (2.17), yields

$$\begin{aligned} \int_{-z_b}^{\zeta} \frac{\partial u_*}{\partial t} \mathrm{d}z + \int_{-z_b}^{\zeta} \frac{\partial {u_*}^2}{\partial x} \mathrm{d}z + \int_{-z_b}^{\zeta} \frac{\partial u_* v_*}{\partial y} \mathrm{d}z + \int_{-z_b}^{\zeta} \frac{\partial u_* w_*}{\partial z} \mathrm{d}z = \int_{-z_b}^{\zeta} f v \, \mathrm{d}z \\ -g \int_{-z_b}^{\zeta} \frac{\partial \zeta}{\partial x} \mathrm{d}z + \frac{1}{\rho} \left[\int_{-z_b}^{\zeta} \frac{\partial \sigma'_{xx}}{\partial x} \mathrm{d}z + \int_{-z_b}^{\zeta} \frac{\partial \tau_{xy}}{\partial y} \mathrm{d}z + \int_{-z_b}^{\zeta} \frac{\partial \tau_{xz}}{\partial z} \mathrm{d}z \right]. \end{aligned} \tag{2.18}$$

Following a similar procedure as above, the application of Leibniz's rule and boundary conditions (2.9) and (2.10), gives

$$\begin{aligned} \frac{\partial (uh)}{\partial t} + \frac{\partial}{\partial x} \int_{-z_b}^{\zeta} u_*^2 \, \mathrm{d}z + \frac{\partial}{\partial y} \int_{-z_b}^{\zeta} u_* v_* \, \mathrm{d}z = f v h - g h \frac{\partial \zeta}{\partial x} \\ + \frac{\tau_{wx} - \tau_{bx}}{\rho} + \frac{1}{\rho} \left[\frac{\partial}{\partial x} \int_{-z_b}^{\zeta} \sigma'_{xx} \, \mathrm{d}z + \frac{\partial}{\partial y} \int_{-z_b}^{\zeta} \tau_{xy} \, \mathrm{d}z \right], \end{aligned} \tag{2.19}$$

where the wind and the bottom friction stress components are defined as

$$\tau_{wx} = \sigma'_{xx} \big|_{z=\zeta} \frac{\partial \zeta}{\partial x} - \tau_{xy} \big|_{z=\zeta} \frac{\partial \zeta}{\partial y} + \tau_{xz} \big|_{z=\zeta}, \tag{2.20}$$

and

$$\tau_{bx} = \sigma'_{xx} \big|_{z=-z_b} \frac{\partial (-z_b)}{\partial x} - \tau_{xy} \big|_{z=-z_b} \frac{\partial (-z_b)}{\partial y} + \tau_{xz} \big|_{z=-z_b}, \tag{2.21}$$

It should be noted that this derivation neglects form drag at the sea bed. A more formal derivation of this may be found in Papadopoulos et al. [7] and Pokrajac [8] Defining

$$\beta_{xx} = \frac{1}{u^2 h} \int_{-z_b}^{\zeta} u_*^2 \, \mathrm{d}z \quad \text{and} \quad \beta_{xy} = \frac{1}{u v h} \int_{-z_b}^{\zeta} u_* v_* \, \mathrm{d}z, \tag{2.22}$$

Equation (2.19) may be recast as

$$\frac{\partial(uh)}{\partial t}+\frac{\partial(\beta_{xx}u^2h)}{\partial x}+\frac{\partial(\beta_{xy}uvh)}{\partial y}=fvh-gh\frac{\partial\zeta}{\partial x}$$
$$+\frac{\tau_{wx}-\tau_{bx}}{\rho}+\frac{1}{\rho}\left[\frac{\partial}{\partial x}\int_{-z_b}^{\zeta}\sigma_{xx}\,\mathrm{d}z+\frac{\partial}{\partial y}\int_{-z_b}^{\zeta}\tau_{xy}\,\mathrm{d}z\right], \tag{2.23}$$

The multipliers β_{xx} and β_{xy} represent correction factors resulting from dispersive horizontal momentum exchanges due to the non-uniformity of the velocity profiles over the depth of the fluid (see Falconer [5] for details). In most practical applications these multipliers are very close to unity and are hence approximated as $\beta_{xx}=\beta_{xy}=1$ in this thesis.

Using the Boussinesq eddy viscosity hypothesis, presented by Boussinesq [3], the Reynolds stresses from (2.6) may be expressed in terms of mean velocity components:

$$\sigma'_{xx}=-\rho\overline{u'_*u'_*}=\rho\nu_t 2\frac{\partial u_*}{\partial x}\quad\text{and}\quad\tau_{xy}=-\rho\overline{u'_*v'_*}=\rho\nu_t\left(\frac{\partial u_*}{\partial y}+\frac{\partial v}{\partial x}\right), \tag{2.24}$$

where ν_t is the turbulent eddy viscosity. The integrals of these terms may be simplified by introducing a depth-scaled turbulent eddy viscosity ν_T which allows rewriting the integrals as

$$\int_{-z_b}^{\zeta}\nu_t\frac{\partial u_*}{\partial x}\,\mathrm{d}z=\nu_T h\frac{\partial u}{\partial x},\quad\int_{-z_b}^{\zeta}\nu_t\frac{\partial u_*}{\partial y}\,\mathrm{d}z=\nu_T h\frac{\partial u}{\partial y},\quad\text{and}$$
$$\int_{-z_b}^{\zeta}\nu_t\frac{\partial v_*}{\partial x}\,\mathrm{d}z=\nu_T h\frac{\partial v}{\partial x}. \tag{2.25}$$

Hence (2.23) may be expressed as

$$\frac{\partial(uh)}{\partial t}+\frac{\partial(\beta_{xx}u^2h)}{\partial x}+\frac{\partial(\beta_{xy}uvh)}{\partial y}=fvh-gh\frac{\partial\zeta}{\partial x}+\frac{\tau_{wx}-\tau_{bx}}{\rho}$$
$$+\frac{1}{\rho}\left[\frac{\partial}{\partial x}\left(2\nu_T h\frac{\partial u}{\partial x}\right)+\frac{\partial}{\partial y}\left(\nu_T h\left(\frac{\partial u}{\partial y}+\frac{\partial v}{\partial x}\right)\right)\right]. \tag{2.26}$$

For large tidal sites where currents vary smoothly over the spatial domain, viscous and turbulence terms may be neglected as they involve second order derivatives in space of depth-averaged velocity. The two-dimensional shallow water equations may then be simplified to

$$\frac{\partial(uh)}{\partial t}+\frac{\partial u^2h}{\partial x}+\frac{\partial uvh}{\partial y}=fvh-gh\frac{\partial\zeta}{\partial x}+\frac{\tau_{wx}-\tau_{bx}}{\rho} \tag{2.27a}$$
$$\frac{\partial(vh)}{\partial t}+\frac{\partial uvh}{\partial x}+\frac{\partial v^2h}{\partial y}=-fuh-gh\frac{\partial\zeta}{\partial y}+\frac{\tau_{wy}-\tau_{by}}{\rho}. \tag{2.27b}$$

Here the left hand side expresses the local and convective acceleration of the flow and the right hand side expresses Coriolis force acceleration, the pressure head driving the flow, and the surface (wind) and bed friction stresses.

For the tidal flow applications considered in Chap. 3, we will consider a reduced form of the above equations, written in one-spatial dimension, and ignoring the terms related to turbulence and wind stress, given by

$$\frac{\partial \zeta}{\partial t} + \frac{\partial (uh)}{\partial x} = 0, \tag{2.28}$$

$$\frac{\partial (uh)}{\partial t} + \frac{\partial (u^2 h)}{\partial x} = -gh\frac{\partial \zeta}{\partial x} - \frac{\tau_{bx}}{\rho}. \tag{2.29}$$

The bed stress is calculated using the following (commonly employed) empirical quadratic approach,

$$\tau_{bx} = \rho C_d \, |u| \, u, \tag{2.30}$$

where C_d is either prescribed, or given by the Chézy or Manning formula by

$$C_d = \sqrt{\frac{g}{C^2}} = \frac{gn^2}{h^{1/3}}, \tag{2.31}$$

where C is Chézy coefficient, and n is the Manning coefficient.

2.3 Conclusions

This chapter describes the derivation of the one-dimensional shallow water equations, in the form of (2.28) and (2.29). These equations constitute the governing equations solved numerically, as described in Chap. 3, for flow in a strait containing a fence of turbines. Further, the shallow water equations are used in Chap. 4 to derive analytic power models for flow through turbine farms modelled as enhanced bed roughness.

References

1. Abbott MB (1979) Computational hydraulics: elements of the theory of free surface flows. Pitman Publishing Limited
2. Adcock TAA, Draper S, Nishino T (2015) Tidal power generation - A review of hydrodynamic modelling. Proc Inst Mech Eng A 1–17
3. Boussinesq J (1877) Essai sur la théorie des eaux courantes. Mémoires présentés par divers savants à l'Académie des Sciences 23:1–680
4. Draper S (2011) Tidal stream energy extraction in coastal basins. PhD thesis, University of Oxford

5. Falconer RA (1993) An introduction to nearly-horizontal flows. In: Abbott MB, Price WA (eds) Coastal, Estuarial and Harbour engineer's reference book. CRC Press, chapter 2, pp 27–36
6. Hughes WF, Brighton JA (1967) Schaum's outline of theory and problems of fluid mechanics. McGraw-Hill, New York
7. Papadopoulos K, Nikora V, Cameron S, Stewart M, Gibbins C (2019) Spatially-averaged flows over mobile rough beds: equations for the second-order velocity moments. J Hydraul Res, (October 2018):1–19
8. Pokrajac D (2013) Depth-integrated Reynolds-averaged Navier-Stokes equations for shallow flows over rough permeable beds. J Hydraul Res 51(5):597–600
9. Reynolds O (1895) On the dynamical theory of incompressible viscous fluids and the determination of the criterion. Phil Trans R Soc Lond A 186:123–164
10. Stansby PK (2006) Limitations of depth-averaged modeling for shallow wakes. J Hydraul Eng 132(7):737–740
11. Taylor GI (1919) Tidal friction in the irish sea. Philos Trans R Soc A: Math Phys Eng Sci 220(571)
12. Vogel CR (2014) Theoretical limits to tidal stream energy extraction. PhD thesis, University of Oxford
13. White FM (2006) Viscous fluid flow. McGraw-Hill, New York

Chapter 3
Uncertain Power from a Hydro-Kinetic Turbine in Steady Flow

3.1 Introduction

This chapter examines the influence of uncertainty in bed friction on hydro-kinetic power extracted from steady flow through a strait represented by a one-dimensional open channel in the stream-wise direction. The flow is driven by a constant head difference between the channel ends. First, an analytic closed-form solution for the power in terms of bed roughness coefficient C_d from a steady one-dimensional model is derived. Next, three methods for the propagation of uncertainty from bed friction to power estimates are compared. The methods are: (1) the transfer of a probability density function (PDF) in C_d through the closed-form solution for power; (2) the numerical equivalent PDF transfer; and (3) an approximate, perturbative method where power is expressed as a Taylor expansion, resulting in leading-order expressions for the statistical moments for power. Further, a numerical model is presented, which is used to determine the expected power and its standard deviation for steady, head-driven flow in a one-dimensional channel. Results from the numerical model are compared to those from the expansion method applied to the closed-form solution to explore the impact of having only an approximate knowledge of the functional dependence of power on bed roughness.

3.2 Analytic Static Channel Model

An analytic model is first derived for the hydro-kinetic power dissipated by a fence of turbines in unidirectional steady flow in a one-dimensional channel driven by a static head difference. Power is expressed in terms of the bed roughness coefficient C_d and turbine drag coefficient C_t. Following the approach of Garrett and Cummins [4] (hereafter GC05), consider an approximation to the one-dimensional shallow water momentum equation (2.29) with bed stress that is quadratic in depth-averaged

M. J. Kreitmair, *The Effect of Uncertainty on Tidal Stream Energy Resource Estimates*, Springer Theses, https://doi.org/10.1007/978-3-030-57658-5_3

flow velocity and an additional term representing momentum dissipation due to the presence of the turbines. In this approach, Eq. (2.29) is amended to include an ad hoc turbine drag term F_{turb}, representing the thrust on a turbine per unit mass of water. For flow in the positive x-direction then, the momentum equation (2.29) may be expressed as

$$\frac{\partial(uh)}{\partial t} + \frac{\partial(u^2 h)}{\partial x} = -gh\frac{\partial\zeta}{\partial x} - C_d u^2 - F_{\text{turb}}, \tag{3.1}$$

where u is streamwise depth-averaged flow velocity, h is local depth, t is time, x is stream-wise distance, g is acceleration due to gravity, ζ is the free surface displacement from still water level, and C_d is the natural bed friction coefficient. Expanding this expression and subtracting the velocity multiplied by the depth-averaged continuity equation, $u(\partial\zeta/\partial t + \partial(uh)/\partial x) = 0$, noting that, for a non-erodible bed, $\partial\zeta/\partial t = \partial h/\partial t$, and dividing through by the water depth h gives

$$\frac{\partial u}{\partial t} + u\frac{\partial u}{\partial x} = -g\frac{\partial\zeta}{\partial x} - \frac{C_d}{h}u^2 - \frac{F_{\text{turb}}}{h}. \tag{3.2}$$

This is equivalent to Eq. (2.1) in GC05. For a channel with a static head difference, the free surface elevation does not vary with time, and the flow rate per unit width $q = uh$ is independent of along-channel distance x and the transient term $\mathrm{d}q/\mathrm{d}t$ disappears. Equation (3.2) may then be recast in terms of the flow rate as

$$\frac{1}{2}q^2\frac{\partial}{\partial x}\left(\frac{1}{h^2}\right) = -g\frac{\partial\zeta}{\partial x} - \frac{C_d}{h^3}q^2 - \frac{F_{\text{turb}}}{h}. \tag{3.3}$$

Integrating this expression with respect to x then gives

$$\frac{1}{2}q^2\left(\frac{1}{h_{\text{out}}^2} - \frac{1}{h_{\text{in}}^2}\right) = -g\left(\zeta_{\text{out}} - \zeta_{\text{in}}\right) - q^2\int_0^L \frac{C_d}{h^3}\,\mathrm{d}x - \int_0^L \frac{F_{\text{turb}}}{h}\,\mathrm{d}x, \tag{3.4}$$

where h_{in} and h_{out} are the water depths at the channel entrance and exit, and ζ_{in} and ζ_{out} are the deviations of the free surface from still water level at channel entrance and exit.

To incorporate the effect of the turbines, it is assumed that they act effectively in the same way as bed roughness, but only over the length of the patch from $x = L_{T1}$ to $x = L_{T2}$, the starting and end locations of the region of effect of the tidal turbines, respectively, a total distance of $L_T = L_{T2} - L_{T1}$. The turbine drag term may therefore be replaced by

$$\int_0^L \frac{F_{\text{turb}}}{h}\,\mathrm{d}x = \int_{L_{T1}}^{L_{T2}} \frac{C_t}{h^3}\,\mathrm{d}x q^2, \tag{3.5}$$

where C_t is the turbine drag coefficient, and Eq. (3.4) is rewritten as

$$-g\,\Delta h = -(\delta_0 + \delta_T)q^2, \tag{3.6}$$

where $\Delta h = \zeta_{\text{in}} - \zeta_{\text{out}}$ is the head difference between the ends of the channel, and the roughness parameters δ_0 and δ_T are defined as

$$\delta_0 = \int_0^L \frac{C_d}{h^3}\,\mathrm{d}x - \frac{1}{2}\left(\frac{1}{h_{\text{out}}^2} - \frac{1}{h_{\text{in}}^2}\right) \tag{3.7}$$

and

$$\delta_T = \int_{L_{T1}}^{L_{T2}} \frac{C_t}{h^3}\,\mathrm{d}x. \tag{3.8}$$

For a static head difference Δh is constant such that the steady state flow rate is given by

$$q = \left(\frac{g\,\Delta h}{\delta_0 + \delta_T}\right)^{1/2}. \tag{3.9}$$

Consequently, the instantaneous power dissipated by the turbines $P_{\text{stat}} = \rho q^3 \delta_T$ becomes

$$P_{\text{stat}} = \rho\,(g\,\Delta h)^{3/2} \frac{\delta_T}{(\delta_0 + \delta_T)^{3/2}}. \tag{3.10}$$

In this static scenario, power has the same functional dependence on bed roughness and turbine drag as for the quasi-steady case with a sinusoidal head difference derived by GC05. The result from GC05 for time-averaged power may be recovered by allowing for a time-varying head difference $\Delta h = a\cos(\omega t)$ and averaging over a tidal period.

Assuming the change in channel depth is small compared to the average depth of the channel, i.e. $\Delta h / h_{avg} \ll 1$, where $h_{avg} = (h_{\text{in}} + h_{\text{out}})/2$, the functions of water depth may be expressed as a series expansion, e.g. $1/h^3 = 1/h_{avg}^3[1 - 3\Delta h/h_{avg} + \mathcal{O}\left((\Delta h/h_{avg})^2\right)]$, and the roughness parameters (3.7) and (3.8) may be expressed, to zeroth order, as

$$\delta_0 = C_d \frac{L}{h_{avg}^3} \quad \text{and} \quad \delta_T = C_t \frac{L_T}{h_{avg}^3}. \tag{3.11}$$

Substituting these expressions into (3.10) gives

$$
\begin{aligned}
P_{\text{stat}} &= \rho\left(g\, \Delta h\, h_{avg}\right)^{3/2} \frac{C_t L_T}{(C_d L + C_t L_T)^{3/2}},\\
&= P_{\text{stat},0} \frac{C_t}{(C_d + C_t L_T/L)^{3/2}},
\end{aligned}
\tag{3.12}
$$

where $P_{\text{stat},0} = \rho\left(g\, \Delta h\, h_{avg}\right)^{3/2} L_T/L^{3/2}$. The power is maximised at a turbine drag coefficient of $C_t^* = 2C_d L/L_T$.

3.3 Uncertainty Propagation Methods

Considered in this thesis are three methods of uncertainty transfer for a power surface defined by a known function of bed roughness coefficient, in terms of expected power and its standard deviation. The first method uses the expression for power to determine analytically the expression for the power PDF from a defined input PDF in C_d, and to use this to calculate statistical moments for power. The second transfers the PDF in C_d numerically to obtain a PDF for power, and uses this to determine expected power $\mathrm{E}[P]$ and the standard deviation in power σ_P. This method may also be used when the exact functional dependence of the power on C_d is unknown, and only a power surface is available (such as would be the case for power determined from a complex numerical model). The third method performs a Taylor expansion on the known expression for power in terms of a small, randomly fluctuating perturbation in C_d, and determines analytic, leading-order expressions for $\mathrm{E}[P]$ and σ_P. These methods are now compared against one another for the power surface in (C_d, C_t) defined by (3.12).

To avoid negative bed roughness values being attributed a finite probability, a truncated normal distribution is used as input distribution for C_d. The truncation sets the probability density for values of $C_d \leq 0$ to zero and, to retain symmetry properties of a normal distribution, was also applied for values $C_d \geq 2\mu_{C_d}$. The PDF for the bed roughness coefficient is therefore given by

$$
f_{C_d}(C_d; \mu_{C_d}, \sigma_{C_d}) = \begin{cases}
0, & \text{for } C_d \leq 0\\
\dfrac{1}{\sqrt{2\pi}\sigma} \dfrac{\exp\left\{-\frac{(C_d-\mu_{C_d})^2}{2\sigma_{C_d}^2}\right\}}{\operatorname{erf}\left(\frac{\mu_{C_d}}{\sqrt{2}\sigma_{C_d}}\right)}, & \text{for } 0 < C_d < 2\mu_{C_d}\\
0, & \text{for } C_d \geq 2\mu_{C_d}.
\end{cases}
\tag{3.13}
$$

Here μ_{C_d} and σ_{C_d} are the mean and standard deviation of the "parent" normal distribution and $\operatorname{erf}(x) = \frac{1}{\sqrt{\pi}} \int_{-x}^{x} e^{-t^2}\, dt$ is the error function. The resulting mean $\mu_{C_d}^*$ and standard deviation $\sigma_{C_d}^*$ of the symmetrically truncated distribution are given by

$$\mu^*_{C_d} = \mu_{C_d} \quad \text{and} \quad \sigma^{*2}_{C_d} = \sigma^2_{C_d}\left(1 - 2\frac{\mu_{C_d}}{\sigma_{C_d}}\frac{1}{\sqrt{2\pi}}\frac{\exp\left\{-\frac{1}{2}\frac{\mu^2_{C_d}}{\sigma^2_{C_d}}\right\}}{\mathrm{erf}\left(\frac{\mu_{C_d}}{\sqrt{2}\sigma_{C_d}}\right)}\right). \tag{3.14}$$

As desired, the mean of the truncated distribution remains the same as that of the parent normal. The standard deviation is reduced due to the cut-off imposed on the tails of the PDF.

3.3.1 Analytic PDF Transfer

The probability density function for power may be determined from standard theory. That is, the PDF $f_P(P)$ for a random variable P, where P is a function of another random variable C_d with known PDF $f_{C_d}(C_d)$, is given by

$$f_P(P) = \sum_i^n \frac{f_{C_d}(C_{d,i})}{\left|\partial P/\partial C_d|_{C_d=C_{d,i}}\right|}, \tag{3.15}$$

where i refers to the ith root of $P(C_d)$ (see Papoulis [11]). Application of Eq. (3.15) to the expression for power (3.12) for a bed roughness coefficient distributed according to the truncated normal distribution defined in (3.13) gives

$$f_P(P) = \begin{cases} 0, & \text{for } P(C_d \le 0) \\ \frac{1}{3}\sqrt{\frac{2}{\pi}}\frac{\exp\left[-\frac{\left(C_t\frac{L_T}{L} + \mu_{C_d} - \left(\frac{1}{C_t}\frac{P}{P_{0,\text{stat}}}\right)^{-2/3}\right)^2}{2\sigma^2_{C_d}}\right]}{\sigma_{C_d} P\left(\frac{1}{C_t}\frac{P}{P_{0,\text{stat}}}\right)^{2/3}}, & \text{for } P(0 < C_d < 2\mu_{C_d}) \\ 0, & \text{for } P(C_d \ge 2\mu_{C_d}). \end{cases} \tag{3.16}$$

This expression is shown plotted against the y-axis in Fig. 3.1, along with corresponding input distributions in C_d plotted against the x-axis, for a range of values of input relative standard deviation.

Expected power may then be determined using

$$\mathrm{E[P]} = \int_{\mathrm{P_{low}}}^{\mathrm{P_{high}}} \mathrm{P\, f_P(P)\, dP}, \tag{3.17}$$

and the variance by

$$\sigma^2_P = \int_{P_{low}}^{P_{high}} (P - \mathrm{E}[P])^2\, f_P(P)\, dP, \tag{3.18}$$

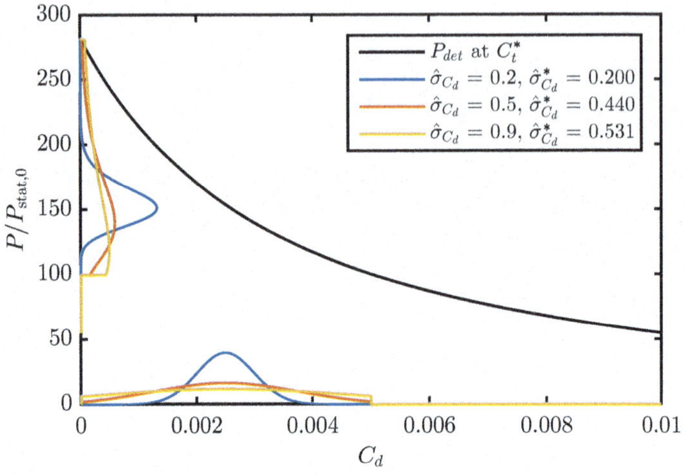

Fig. 3.1 Transfer of PDFs in C_d (plotted along x-axis) through the deterministic power curve at $C_d = \mu_{C_d} = 0.0025$ and $C_t = C_t^*$ to the corresponding PDFs for power (y-axis), for truncated normal input distributions with standard deviation values $\hat{\sigma}^*_{C_d} = \{0.200, 0.440, 0.531\}$. The heights of the distributions have been scaled for ease of illustration

where P_{low} and P_{high} are the lower and upper bounds of the truncated PDF in P given by (3.12) evaluated at $C_d = 2\mu_{C_d}$ and $C_d = 0$, respectively.

3.3.2 Numerical PDF Transfer

In situations where the functional dependence of power on bed roughness coefficient is not known, it is useful to transfer the input PDF to the associated PDF for power via numerical means, i.e. discretely transferring the probability of a value in C_d through the power surface to the associated power value. The mathematics for this method is first expressed below, followed by the detailed procedure. This is followed by a comparison of the power PDFs produced using the analytic PDF transfer from the previous subsection and the numerical transfer.

The probability of a given value for bed roughness coefficient is assumed to follow the normal distribution within the truncated region, $\mathcal{N}$ with mean μ_{C_d} and variance ${\sigma_{C_d}}^2$,

$$\mathcal{N}(C_d \mid \mu_{C_d}, {\sigma_{C_d}}^2) = \frac{1}{\sqrt{2\pi {\sigma_{C_d}}^2}} e^{-\frac{(C_d - \mu_{C_d})^2}{2{\sigma_{C_d}}^2}}. \tag{3.19}$$

The corresponding cumulative distribution function (CDF), i.e. the probability that the random variable C_d will take a value smaller than $C_{d,0}$, for a normally distributed process is given by

$$\Phi(C_{d,0}) = \frac{1}{\sqrt{2\pi\sigma_{C_d}^2}} \int_{-\infty}^{C_{d,0}} e^{-\frac{(C_d' - \mu_{C_d})^2}{2\sigma_{C_d}{}^2}} \mathrm{d}C_d' \tag{3.20}$$

$$= \frac{1}{2}\left[1 + \mathrm{erf}\left(\frac{C_{d,0} - \mu_{C_d}}{\sigma_{C_d}\sqrt{2}}\right)\right]. \tag{3.21}$$

Hence, the probability that the value of a realisation of the random variable C_d falls between the values $C_{d,A}$ and $C_{d,B}$ (where $C_{d,A} < C_{d,B}$) is given by

$$\begin{aligned} Pr(C_{d,A} \leq C_d < C_{d,B}) &= \int_{C_{d,A}}^{C_{d,B}} \mathcal{N}(C_d \mid \mu_{C_d}, \sigma_{C_d}^2)\,\mathrm{d}C_d \\ &= \int_{-\infty}^{C_{d,B}} \mathcal{N}(C_d \mid \mu_{C_d}, \sigma_{C_d}^2)\,\mathrm{d}C_d - \int_{-\infty}^{C_{d,A}} \mathcal{N}(C_d \mid \mu_{C_d}, \sigma_{C_d}^2)\,\mathrm{d}C_d \\ &= \Phi(C_{d,B}) - \Phi(C_{d,A}) \\ &= \frac{1}{2}\left[\mathrm{erf}\left(\frac{C_{d,B} - \mu_{C_d}}{\sigma_{C_d}\sqrt{2}}\right) - \mathrm{erf}\left(\frac{C_{d,A} - \mu_{C_d}}{\sigma_{C_d}\sqrt{2}}\right)\right]. \end{aligned} \tag{3.22}$$

For sufficiently finely spaced C_d values, the likelihood of a value of $C_{d,i}$ being realised may be defined as the probability of the random variable falling within the interval bounded by $C_{d,i-1/2}$ and $C_{d,i+1/2}$. Using (3.22), this is given by

$$\begin{aligned} Pr(C_d = C_{d,i}) &= Pr(C_{d,i-1/2} \leq C_d < C_{d,i+1/2}) \\ &= Pr\left(\frac{1}{2}(C_{d,i-1} + C_{d,i}) \leq C_d < \frac{1}{2}(C_{d,i} + C_{d,i+1})\right) \\ &= \frac{1}{2}\left[\mathrm{erf}\left(\frac{C_{d,i+1/2} - \mu_{C_d}}{\sigma_{C_d}\sqrt{2}}\right) - \mathrm{erf}\left(\frac{C_{d,i-1/2} - \mu_{C_d}}{\sigma_{C_d}\sqrt{2}}\right)\right]. \end{aligned} \tag{3.23}$$

This may then be used to calculate the expected power for a given value of C_t using

$$E[P] = \sum^{i} P_i(C_d = C_{d,i}) Pr(C_d = C_{d,i}) \tag{3.24}$$

and the variance as a function of C_t using

$$\sigma_P^2 = \sum^{i} (P_i(C_d = C_{d,i}) - E[P])^2 Pr(C_d = C_{d,i}). \tag{3.25}$$

In summary, the procedure for calculating the statistical moments for power P from a given distribution of bed roughness coefficient C_d is as follows:

- Create a computational model of one-dimensional flow in a strait spanned by a fence of turbines where the bed friction is C_d and the turbine drag is C_t.
- Run the model systematically for an array of different C_d and C_t values, and record the power output in a look-up table.
- Use two-dimensional cubic spline to refine the look-up table.
- Define a probability density function for C_d with specified mean μ_{C_d} and variance $\sigma^2_{C_d}$.
- For each value of turbine drag C_t, subdivide the PDF into bins (see Fig. 3.2), and calculate the probability of C_d within each bin by integrating over the width of each bin. This gives the associated probability of P by conservation of probability.
- Having determined the probability associated with a given value of power $P = f(C_d)$, then determine the expected value using

$$E[P] = \sum_i P_i(C_d = C_{d,i})\text{Pr}(C_d = C_{d,i}), \tag{3.26}$$

and the variance

$$\sigma_P^2 = \sum_i (P_i(C_d = C_{d,i}) - E[P])^2\text{Pr}(C_d = C_{d,i}). \tag{3.27}$$

Higher, nth order statistical moments are given by

$$\mu_n = \sum_i (P_i(C_d = C_{d,i}) - E[P])^n\text{Pr}(C_d = C_{d,i}), \tag{3.28}$$

where $n = 3$ gives the skewness of a distribution (a measure for asymmetry) and μ_4 gives the kurtosis (a measure for the "peakedness" or "tailedness").

The computations carried out in this thesis were performed using MATLAB® software [10].

Figure 3.3a shows the probability density distributions for C_d and Fig. 3.3b the corresponding PDFs for power generated from the analytic expressions (3.13) and (3.16) and using the numerical PDF transfer approach outlined above. There is near perfect agreement between the results from the two methods for PDF transfer, with tiny differences occurring due to discretisation in the numerical transfer. This indicates that use of the numerical PDF transfer method is appropriate for a sufficiently finely resolved C_d grid. Consequently, the results determined from the numerical transfer method may be viewed as sufficiently accurate for comparison with the expansion method which is outlined next.

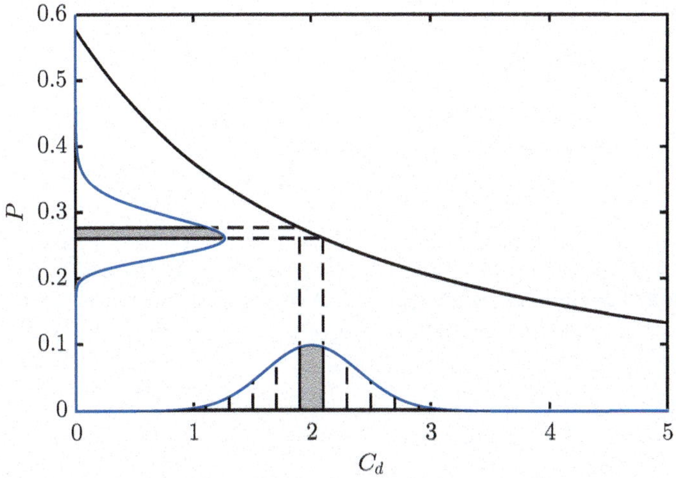

Fig. 3.2 Probability density transfer from a PDF for C_d to a PDF in P via a function $P = f(C_d)$. The shaded regions represent the same area, by conservation of probability

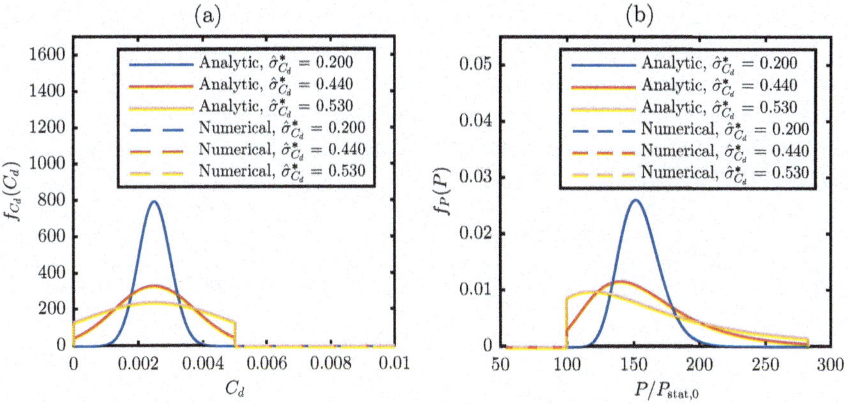

Fig. 3.3 **a** Input probability distribution function for C_d (truncated normal); and **b** corresponding PDF for power at $C_d = \mu_{C_d} = 0.0025$ and $C_t = C_t^*$ for $\hat{\sigma}^*_{C_d} = \{0.200, 0.440, 0.531\}$

3.3.3 Expansion Method

Uncertainty in C_d is introduced by expressing the coefficient in Eq. (3.12) as a random variable with mean μ_{C_d}, i.e. $C_d = \mu_{C_d} + \Delta C_d$, where ΔC_d is a zero-mean fluctuation. For a sufficiently small ΔC_d, the power (3.12) may be expanded as a Taylor series in ΔC_d about the deterministic case of $\Delta C_d = 0$:

$$\frac{1}{P_{\mathrm{stat},0}} P_{\mathrm{stat}} = \frac{C_t}{(\mu_{C_d} + C_t \frac{L_T}{L})^{3/2}} - \frac{3}{2} \frac{C_t}{(\mu_{C_d} + C_t \frac{L_T}{L})^{5/2}} \Delta C_d + \frac{15}{8} \frac{C_t}{(\mu_{C_d} + C_t \frac{L_T}{L})^{7/2}} \Delta C_d{}^2 - \frac{35}{16} \frac{C_t}{(\mu_{C_d} + C_t \frac{L_T}{L})^{9/2}} \Delta C_d{}^3 + \frac{315}{128} \frac{C_t}{(\mu_{C_d} + C_t \frac{L_T}{L})^{11/2}} \Delta C_d{}^4 + \mathcal{O}\left(\Delta\lambda_0{}^5\right). \tag{3.29}$$

The series converges for sufficiently small ΔC_d, i.e. $\Delta C_d/\mu_{C_d} \ll 1$. Applying the expectation operator to the expansion (3.29), the expected power for an uncertain C_d with a mean μ_{C_d} and a standard deviation of σ_{C_d} is given by

$$\frac{\mathrm{E}[P_{\mathrm{stat}}]}{P_{\mathrm{stat},0}} = \frac{C_t}{(\mu_{C_d} + C_t \frac{L_T}{L})^{3/2}} + \frac{15}{8} \frac{C_t}{(\mu_{C_d} + C_t \frac{L_T}{L})^{7/2}} \sigma_{C_d}^2 - \frac{35}{16} \frac{C_t}{(\mu_{C_d} + C_t \frac{L_T}{L})^{9/2}} \mu_{C_d,3} + \frac{315}{128} \frac{C_t}{(\mu_{C_d} + C_t \frac{L_T}{L})^{11/2}} \mu_{C_d,4} + \mathcal{O}\left(\mathrm{E}[\Delta C_d{}^5]\right), \tag{3.30}$$

where $\mu_{C_d,3} = \mathrm{E}[\Delta C_d{}^3]$ is the skewness, and $\mu_{C_d,4} = \mathrm{E}[\Delta C_d{}^4]$ is the kurtosis of the input distribution for C_d. The first term corresponds to the deterministic power at $C_d = \mu_{C_d}$ while the second term represents a stochastic correction resulting from the spread of C_d values about the mean, represented by the variance $\sigma_{C_d}^2$. Higher order terms take into account asymmetry and "tailedness" of the input PDF.

Similarly, the standard deviation in power is found by evaluating $\sigma_P{}^2 = \mathrm{E}\left[(P - \mathrm{E}[P])^2\right]$, giving

$$\frac{\sigma_P^2}{P_{\mathrm{stat},0}^2} = \frac{9}{4} \frac{C_t^2}{(\mu_{C_d} + C_t \frac{L_T}{L})^5} \sigma_{C_d}^2 - \frac{45}{8} \frac{C_t^2}{(\mu_{C_d} + C_t \frac{L_T}{L})^6} \mu_{C_d,3} + \frac{645}{64} \frac{C_t^2}{(\mu_{C_d} + C_t \frac{L_T}{L})^7} \sigma_{C_d}^4 - \frac{225}{64} \frac{C_t^2}{(\mu_{C_d} + C_t \frac{L_T}{L})^7} \mu_{C_d,4} + \mathcal{O}\left(\mathrm{E}[\Delta C_d{}^3]\right). \tag{3.31}$$

To test the effectiveness of the expansion method used to derive the expressions (3.30) and (3.31), the results were compared to values of expected power and standard deviation determined by using the numerical PDF transfer method above, at increasing values of standard deviation in bed roughness coefficient to explore how the convergence of the series is affected by increasing uncertainty.

Figure 3.4 shows a comparison of the expected power and standard deviation determined from the expansion method at 2nd and 4th order accuracy, against that calculated from the PDF transfer method. As $\hat{\sigma}_{C_d}^*$ increases, expected power is still very well approximated by the 2nd order expansion, with the increase to 4th order improving the estimate only marginally even at a significant input distribution spread of $\hat{\sigma}_{C_d}^* = 0.53$ (Fig. 3.4e).

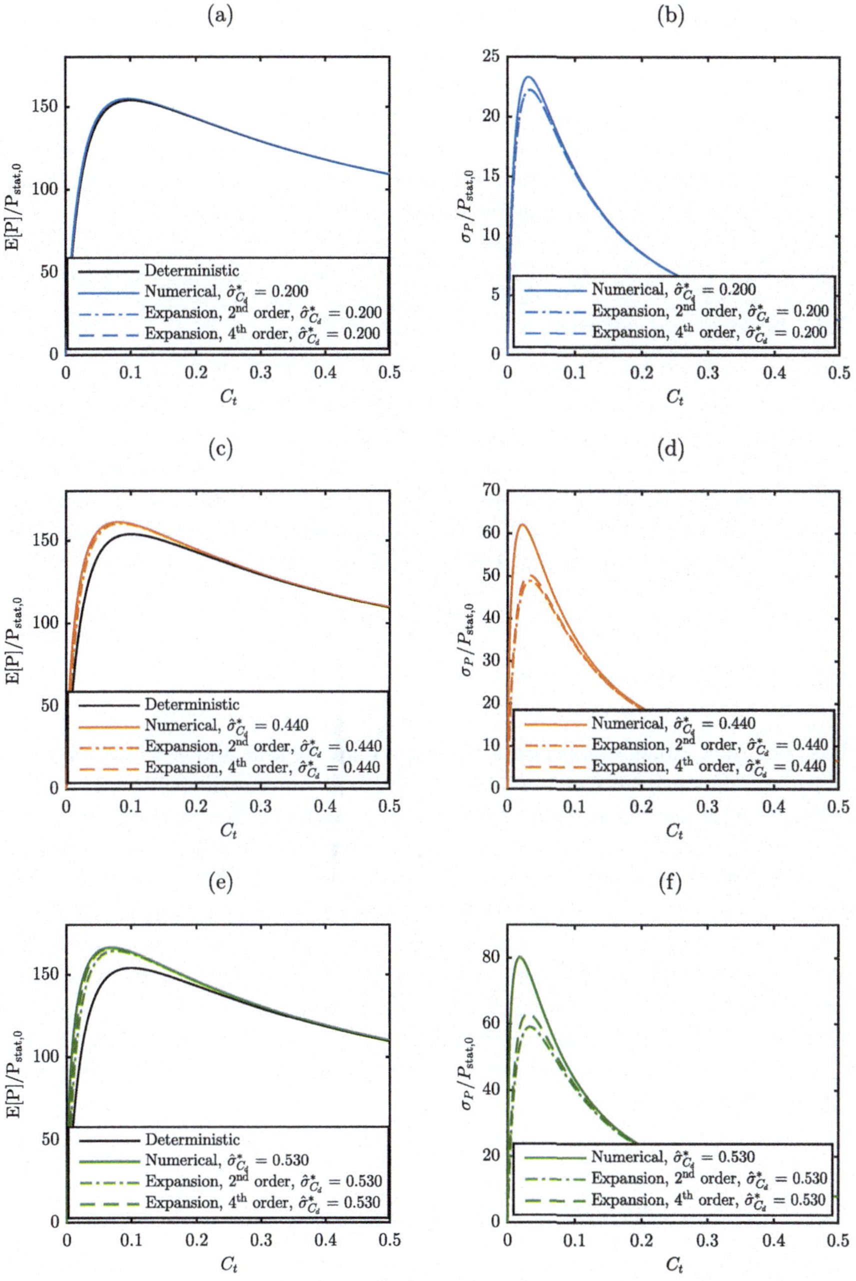

Fig. 3.4 Expected power (left-hand panels) and standard deviation in power (right-hand panels) for the static channel model at input PDF standard deviation $\hat{\sigma}^*_{C_d} = \{0.200, 0.440, 0.531\}$

Standard deviation in power is more sensitive to the accuracy of the expansion. While underestimating the standard deviation for power at 2nd order, the estimate of the expansion is improved by extension to the 4th order.

This shows, provided that power surface is sufficiently well captured by a closed form solution, that the expansion technique set out in this section gives a good, leading-order estimate for the effect of bed roughness uncertainty on power estimates though a slightly less accurate estimate of the standard deviation.

3.4 Numerical Model

In this section a numerical model is presented which is used to calculate numerically the power extracted by turbines spanning a one-dimensional, head-driven channel. The numerical model is then used to compare the expected power and standard deviation determined using the power surface populated from runs of the numerical model at different values of bed roughness coefficient and turbine drag parameter, to those obtained using Eqs. (3.30) and (3.31), respectively.

3.4.1 Finite Volume Solver

The numerical model comprises a Godunov finite volume shock-capturing solver of the one-dimensional shallow water equations in conservative form. The solver is based on an integral version of the governing partial differential equations, and so performs reliably in the presence of discontinuities, such as a step change in bed roughness experienced by flow due to the presence of an energy extraction device. A balanced stage-discharge formulation of the shallow water equations is chosen, following [9], because it utilises a fixed horizontal datum when evaluating free surface flow behaviour and thus avoids generation of unphysical fluxes over non-uniform bathymetry (see [12, 13] for an explanation of such fluxes). The numerical model uses a Harten Lax van Leer contact scheme to solve the hyperbolic equation system, with MUSCL-Hancock second-order time integration. A brief outline of the methodology is given below.

3.4.2 Balanced Shallow Water Equations

The one-dimensional shallow water equations (SWEs) are given in conservative form by the mass and momentum conservation equations:

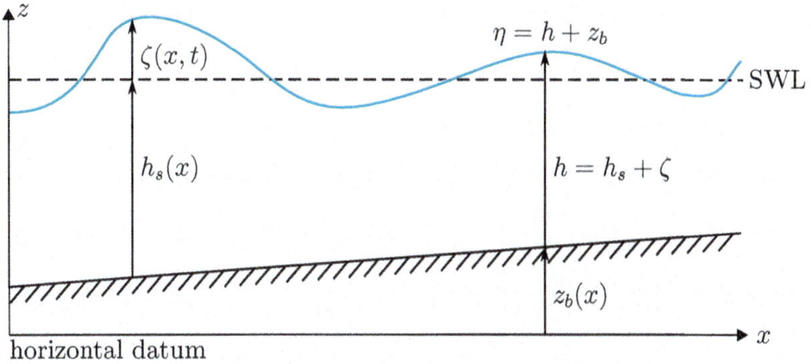

Fig. 3.5 Definition sketch for the stage-discharge formulation of the shallow water equations

$$\frac{\partial h}{\partial t} + \frac{\partial (uh)}{\partial x} = 0, \tag{3.32a}$$

$$\frac{\partial (uh)}{\partial t} + \frac{\partial (u^2 h)}{\partial x} = -gh\frac{\partial \zeta}{\partial x} - \frac{\tau_b}{\rho}, \tag{3.32b}$$

where h is the flow depth, u is the depth-averaged flow velocity, ζ is the deviation of the free surface from the still water level (see Fig. 3.5), x is the horizontal distance, t is time, g is the acceleration due to gravity, and ρ is water density. The bed stress is given by $\tau_b = \rho C_d u\,|u|$, where C_d is the natural bed roughness coefficient. By splitting the free surface gradient source term in (3.32b) into flux and bed gradient components, i.e.

$$gh\frac{\partial \zeta}{\partial x} = \frac{\partial}{\partial x}(\frac{1}{2}gh^2) + gh\frac{\partial z_b}{\partial x}, \tag{3.33}$$

where z_b is the bed elevation above a fixed horizontal datum, the system of conservation laws (3.32a) and (3.32b) can be rewritten in hyperbolic form[1] so that they can be solved numerically using a Godunov-type approximate Riemann scheme (see [15]). However, this formulation is unbalanced when solved within a Godunov-type scheme because non-physical numerical fluxes are calculated and the resulting solutions are not physically meaningful (see [12, 13]).

The present solver is based on a balanced stage-discharge (η, q) formulation, where η is the free surface elevation measured from a horizontal datum and q is the flow rate per unit width, of the hyperbolic form that caters for multiple water bodies with different still water levels as originally proposed by Liang and Borthwick [9].

[1] An $m \times m$ system of partial differential equations of the form of a conservation law $\partial_t \mathbf{q} + \partial_x \mathbf{f} = \mathbf{s}$ may be classified by considering the quasi-linear form $\partial_t \mathbf{q} + \mathbf{A}\partial_x \mathbf{q} = \mathbf{s}$. Here $\mathbf{A}$ is the Jacobian matrix $\mathbf{A} = \partial_{\mathbf{q}} \mathbf{f}$, with eigenvalues λ_i, $i = 1, 2, \ldots, m$. A 2×2 system with two distinct and real eigenvalues is strictly hyperbolic and has two linearly independent eigenvectors defining the characteristics of the system (see [16]).

Here the free surface gradient term is split as follows

$$gh\frac{\partial \zeta}{\partial x} = \frac{\partial}{\partial x}\left(\frac{1}{2}g(\eta^2 - 2\eta z_b)\right) + g\eta\frac{\partial z_b}{\partial x}, \tag{3.34}$$

and, using the relation $\partial h/\partial t = \partial \eta/\partial t$ for a fixed bed, the following system of partial differential equations is obtained

$$\frac{\partial \eta}{\partial t} + \frac{\partial (uh)}{\partial x} = 0, \tag{3.35a}$$

$$\frac{\partial (uh)}{\partial t} + \frac{\partial}{\partial x}\left(u^2 h + \frac{1}{2}g(\eta^2 - 2\eta z_b)\right) = -g\eta\frac{\partial z_b}{\partial x} - \frac{\tau_b}{\rho}. \tag{3.35b}$$

This formulation is discretised using the following finite volume Godunov-type scheme.

3.4.3 Godunov-Type Scheme

In order to ensure conservation of mass and momentum, a finite volume discretisation scheme is used which integrates the balanced shallow water equations over control volumes which tile the computational domain. For steep-fronted or transcritical flows, the flow discontinuity is modelled using a Godunov-type scheme, where, following [5], the continuous flow is discretised onto a grid leading to a series of piece wise continuous data states connected by discontinuities. Each of these is equivalent to a Riemann problem: an initial value problem consisting of conservation equations with constant data states connected by a single discontinuity. In the finite volume scheme, an approximate Riemann solver is invoked to solve this problem as follows.

First, the balanced one-dimensional shallow water Eq. (3.35) in stage-discharge form are recast into a hyperbolic system of partial differential equations given by

$$\frac{\partial \mathbf{q}}{\partial t} + \frac{\partial \mathbf{f}(\mathbf{q})}{\partial x} = \mathbf{s}, \tag{3.36}$$

where

$$\mathbf{q} = \begin{pmatrix} \eta \\ uh \end{pmatrix}, \quad \mathbf{f} = \begin{pmatrix} uh \\ u^2 h + \frac{1}{2}g(\eta^2 - 2\eta z_b) \end{pmatrix}, \quad \text{and} \quad \mathbf{s} = \begin{pmatrix} 0 \\ -g\eta\frac{\partial z_b}{\partial x} - \frac{\tau_b}{\rho} \end{pmatrix}. \tag{3.37}$$

By integrating over a control volume Ω, the conservation law (3.36) may be expressed in integral form as

$$\frac{\partial}{\partial t}\int_{\Omega}\mathbf{q}\,\mathrm{d}\Omega+\int_{\Omega}\frac{\partial \mathbf{f}}{\partial x}\,\mathrm{d}\Omega=\int_{\Omega}\mathbf{s}\,\mathrm{d}\Omega. \tag{3.38}$$

Application of Gauss' theorem to the flux term gives

$$\frac{\partial}{\partial t}\int_{\Omega}\mathbf{q}\,\mathrm{d}\Omega+\oint_{S}\mathbf{f}\,\mathrm{d}S=\int_{\Omega}\mathbf{s}\,\mathrm{d}\Omega, \tag{3.39}$$

where S is a surface bounding the control volume Ω. Consequently, for the ith grid cell the discretised conservation law may be written as

$$\left.\frac{\partial V\mathbf{q}}{\partial t}\right|_i=-\oint_{\partial C_i}\mathbf{f}_i\,\mathrm{d}S+V\mathbf{s}_i=-(\mathbf{f}_{i+1/2}-\mathbf{f}_{i-1/2})+V\mathbf{s}_i, \tag{3.40}$$

where V is the volume of the ith cell. Explicitly for the discretised grid, the formula for the data state $\mathbf{q}$ in cell i at the $(n+1)$th time step is given by

$$\mathbf{q}_i^{n+1}=\mathbf{q}_i^n+\frac{\Delta t}{\Delta x}(\mathbf{f}_{i-1/2}^n-\mathbf{f}_{i+1/2}^n)+\Delta t\,\mathbf{s}_i, \tag{3.41}$$

where Δx is the length of the cell and Δt is the time step. In other words, to determine the updated data state $\mathbf{q}_i^{n+1}$ the fluxes across the cell boundaries $\mathbf{f}_{i\pm1/2}^n$ need to be computed, as shown in Fig. 3.6. This is done using an HLLC solver for the Riemann problem generated at each cell interface.

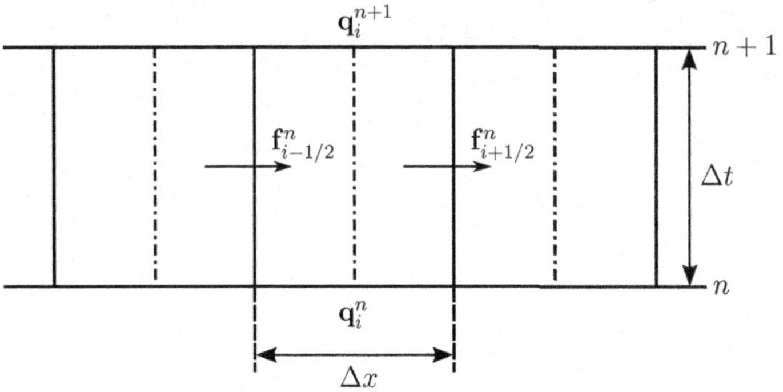

Fig. 3.6 One-dimensional computational mesh in the $x-t$ plane. Fluxes leaving one cell and entering the adjacent cell are identical, ensuring the scheme is conservative. Adapted from Toro [14]

3.4.4 The HLLC Solver

A Harten-Lax-van Leer with contact wave (HLLC) solver developed by Toro et al. [17] is used to determine the inter-cell fluxes $\mathbf{f}_{i\pm1/2}$. It should be noted that a contact, or intermediate, wave can arise in the solution of the Riemann problem posed for the shallow water equations in scenarios with discontinuities or 2D flow over a dry bed. Before outlining the numerical procedure, it is worth pausing to consider the Riemann problem itself.

The Riemann Problem

The one-dimensional Riemann problem involves solving a conservation law (neglecting source terms), written in general differential form as

$$\frac{\partial \mathbf{q}}{\partial t} + \frac{\partial \mathbf{f}(\mathbf{q})}{\partial x} = 0, \tag{3.42}$$

where $\mathbf{q}$ is a vector of (otherwise continuous) conserved flow variables, and $\mathbf{f}(\mathbf{q})$ a vector of the corresponding fluxes, at a discontinuity of the form

$$\mathbf{q} = \begin{cases} \mathbf{q}_L & \text{for } x \leq 0, \\ \mathbf{q}_R & \text{for } x > 0, \end{cases} \tag{3.43}$$

where $\mathbf{q}_L$ and $\mathbf{q}_R$ are the vectors of conserved flow variables either side of the discontinuity interface, as illustrated in Fig. 3.7a.

The conservation law (3.42) may be expressed in terms of the flux Jacobian $\mathbf{A} = \partial\mathbf{f}/\partial\mathbf{q}$ as

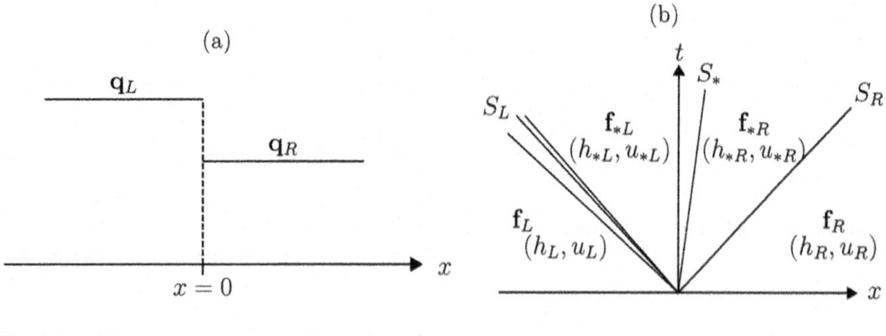

Fig. 3.7 **a** Setup of a Riemann problem with left and right data states are separated by a discontinuity at $x = 0$; and **b** HLLC solution structure of a Riemann problem. Adapted from Liang and Borthwick [9]

$$\frac{\partial \mathbf{q}}{\partial t} + \mathbf{A}(\mathbf{q})\frac{\partial \mathbf{q}}{\partial x} = 0. \tag{3.44}$$

This is of the form of a wave equation, with a wave structure solution where the characteristic solutions propagate at wave speeds S_L, S_R, and S_*, as shown in Fig. 3.7b. Provided these wave speeds are known, then the numerical fluxes can be calculated using

$$\mathbf{f}_{i+1/2} = \begin{cases} \mathbf{f}_L & \text{if } 0 \leq S_L, \\ \mathbf{f}_{*L} & \text{if } S_L \leq 0 \leq S_*, \\ \mathbf{f}_{*R} & \text{if } S_* \leq 0 \leq S_R, \\ \mathbf{f}_R & \text{if } 0 \geq S_R, \end{cases} \tag{3.45}$$

where

$$\mathbf{f}_{*L} = \mathbf{f}_L + S_L(\mathbf{q}_{*L} - \mathbf{q}_L), \tag{3.46}$$
$$\mathbf{f}_{*R} = \mathbf{f}_R + S_R(\mathbf{q}_{*R} - \mathbf{q}_R) \tag{3.47}$$

The right and left wave speeds for the two-wave Riemann problem are given by Fraccarollo and Toro [3] as

$$S_L = \begin{cases} u_R - 2\sqrt{gh_R} & \text{if } h_{\text{out}} = 0, \\ \min(u_L - \sqrt{gh_{\text{out}}}, u_* + \sqrt{gh_*}) & \text{if } h_{\text{out}} > 0, \end{cases} \tag{3.48}$$

and

$$S_R = \begin{cases} u_L + 2\sqrt{gh_{\text{out}}} & \text{if } h_R = 0, \\ \max(u_R + \sqrt{gh_R}, u_* + \sqrt{gh_*}) & \text{if } h_R > 0, \end{cases} \tag{3.49}$$

where $u_{L,R}$ and $h_{L,R}$ are the flow velocity and depth of the left and right constant Riemann states of the initial boundary value problem, see Fig. 3.7a. The flow velocity and depth in the intermediate (star) region are given by

$$u_* = \frac{1}{2}(u_L + U_R) + \sqrt{gh_{\text{out}}} - \sqrt{hh_R}, \tag{3.50}$$

and

$$h_* = \frac{1}{g}\left[\frac{1}{2}(\sqrt{gh_{\text{out}}} + \sqrt{hh_R}) + \frac{1}{4}(u_L - u_R)\right]^2. \tag{3.51}$$

For dry-bed problems, [15] suggests calculating the intermediate wave speed using

$$S_* = \frac{S_L h_R(u_R - S_R) - S_R h_{\text{out}}(u_L - S_L)}{h_R(u_R - S_R) - h_{\text{out}}(u_L - S_L)}. \tag{3.52}$$

This constitutes the heart of the HLLC solver, whereby Eqs. (3.45)–(3.52) are computed at each cell interface. In the numerical model, the vectors $\mathbf{q}$ and $\mathbf{f}$ are given in (3.37). Without further modification, the HLLC scheme is formally first order accurate in time but may be improved to second order by implementing a MUSCL-Hancock scheme which is outlined below.

3.4.5 *MUSCL-Hancock Scheme*

The MUSCL-Hancock approximate Riemann solver is a predictor-corrector approach which improves the accuracy of a Godunov-type scheme to second order by replacing the piece-wise constant approximation of the cell-states with states reconstructed from a previous time step. The method consists of three main steps:

(i) Reconstruction In the first step, the cell-averaged data states are reconstructed as piece-wise linear functions through interpolation. As a result, the interpolated data states at the west and east cell interfaces, denoted by subscripts W and E respectively, are given by

$$\mathbf{q}^n_{i,\text{W}} = \mathbf{q}^n_i - \frac{1}{2}(\mathbf{q}^n_i - \mathbf{q}^n_{i-1}) \quad \text{and} \quad \mathbf{q}^n_{i,\text{E}} = \mathbf{q}^n_i + \frac{1}{2}(\mathbf{q}^n_i - \mathbf{q}^n_{i-1}). \tag{3.53}$$

The piece-wise linear functions satisfy conservation in that they may be averaged over the cell to give the piece-wise constant value that they replace.

Slope Limiting Application of a slope-limiter $\Phi(\mathbf{r})$ to the reconstructed data states prevents the production of spurious oscillations which occur in high-order schemes in regions of steep gradients. Using this, the reconstruction given in (3.53) is changed to

$$\mathbf{q}^n_{i,E/W} = \mathbf{q}^n_i \pm \frac{1}{2}\Phi(\mathbf{r})(\mathbf{q}^n_i - \mathbf{q}^n_{i-1}) \tag{3.54}$$

where $\mathbf{r}$ is the ratio of neighbouring gradients:

$$\mathbf{r} = \begin{cases} \dfrac{\mathbf{q}_{i+1} - \mathbf{q}_i}{\mathbf{q}_i - \mathbf{q}_{i-1}} & \text{if } \mathbf{q}_i \neq \mathbf{q}_{i-1}, \\ 0 & \text{if } \mathbf{q}_i = \mathbf{q}_{i-1} \end{cases} \tag{3.55}$$

The value of the slope limiter is given by

$$\Phi(\mathbf{r}) = \max[0, \min(\beta, \mathbf{r}), \min(1, \beta\mathbf{r})], \tag{3.56}$$

which corresponds to the MINMOD limiter for $\beta = 1$ and the SUPERBEE for $\beta = 2$, as described by Hirsch [8].

(ii) Predictor Step The reconstructed states at the cell interfaces are evolved forward in time by half a time step $\Delta t/2$ using

$$\mathbf{q}_{i,E/W}^{n+1/2} = \mathbf{q}_{i,E/W}^{n} - \frac{1}{2}\frac{\Delta t}{\Delta x}\left(\mathbf{f}(\mathbf{q}_{i,E}) - \mathbf{f}(\mathbf{q}_{i,W})\right) + \frac{1}{2}\Delta t\, \mathbf{s}_i. \quad (3.57)$$

(iii) Corrector Step Finally, the predicted states at the interface are utilised as initial data states of a conventional Riemann problem, using the HLLC solver described above.

3.4.6 Boundary Conditions

At the ends of the numerical domain, boundary conditions need to be imposed to allow the numerical solver to access updated data beyond the computational domain, which may be necessary to calculate gradients and fluxes at the edges of the first and last cell. Boundary conditions used in this thesis are predominantly extrapolated, transmissive or prescribed.

Extrapolated and Transmissive

Extrapolated and transmissive boundary conditions allow flow into and out of the domain, by extrapolating the relevant variable. For a domain spanning from $i = 1$ to $i = i_{max}$, the end boundary conditions are given by

$$\eta_0^{n+1} = 2\eta_1^{n+1} - \eta_2^{n+1}; \qquad \eta_{-1}^{n+1} = 2\eta_0^{n+1} - \eta_1^{n+1} \quad (3.58)$$

and

$$\eta_{i_{max}+1}^{n+1} = 2\eta_{i_{max}}^{n+1} - \eta_{i_{max}-1}^{n+1}; \qquad \eta_{i_{max}+2}^{n+1} = 2\eta_{i_{max}+1}^{n+1} - \eta_{i_{max}}^{n+1} \quad (3.59)$$

for the extrapolated water depth at the inlet and outlet of the channel, and by

$$q_0^{n+1} = q_1^{n+1}; \qquad q_{-1}^{n+1} = q_0^{n+1} \text{ and } q_{i_{max}+1}^{n+1} = q_{i_{max}}^{n+1}; \qquad q_{i_{max}+2}^{n+1} = q_{i_{max}+1}^{n+1} \quad (3.60)$$

for the transmissive flow rate.

Prescribed

Prescribed, or clamped, boundary conditions specify the water depth or the flow rate at the inlet or outlet, i.e.

$$\eta_0^{n+1} = \eta_0^{n+}; \qquad \eta_{-1}^{n+1} = \eta_{-1}^{n} \text{ and } \eta_{i_{max}}^{n+1} = \eta_{i_{max}}^{n}; \qquad \eta_{i_{max}+1}^{n+1} = \eta_{i_{max}+1}^{n} \tag{3.61}$$

for the free surface elevation and equivalently for discharge.

3.4.7 Model Verification Tests

The shallow water solver was verified against several test cases with either analytic solutions or predictions available from very high resolution numerical models (such as that of Toro [15]). The cases test the ability of the solver to conserve mass and momentum. The first four tests, cases 1–4, are performed within a domain of length $x_L = 50$ m, with a flat and frictionless bed, and transmissive boundary conditions at both ends of the domain. After convergence testing, the numerical parameters of these four cases were set to $\Delta t = 0.001$ s for the time step and to $\Delta x = 0.0625$ m for the grid spacing. The fifth and final test case, transcritical flow over a hump, is described in more detail below.

Case 1 - Left Critical Rarefaction and Right Shock

This dam break problem over a frictionless, wet bed tests the ability of the solver to capture adequately a right-propagating shock and the corresponding left-propagating rarefaction wave. Initial conditions are given by water depths of $h_{\text{out}} = 1.0$ m and $h_R = 0.1$ m to the left and right, respectively, of the location of the discontinuity at $x = 10$ m. The corresponding flow velocities are $u_L = 2.5$ ms^{-1} and $u_R = 0.0$ ms^{-1}. Figure 3.8 shows the free surface elevation and flow velocity profiles along the channel at initial time t = 0.0 s and final time t = 7.0 s. The final profiles show that a bore has developed with an almost vertical front (both in depth and velocity), there is a plateau region downstream of the front, and an almost linearly changing rarefaction wave further downstream. Neither the bore nor the rarefaction have reached the ends of the domain. The profiles are in extremely close agreement with those obtained by Toro [15] and others. The present solver predicts a depth $h_* = 0.6116$ m and flow velocity $u_* = 3.865$ ms^{-1} within the star region, which compare well to the very high resolution values of $h_* = 0.611753$ m and $u_* = 3.86398$ ms^{-1} obtained by Toro [15]. The reverse simulation where a left-propagating bore was produced, gave exactly mirror image results, as would be expected.

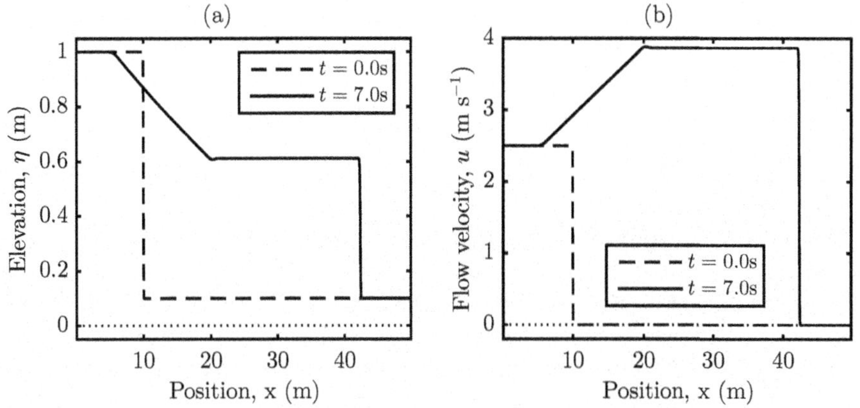

Fig. 3.8 Case 1: left critical rarefaction and right shock at initial conditions and $t = 7.0$ s. **a** Free surface elevation profiles; **b** flow velocity profiles

Case 2 - Two Rarefactions and Nearly Dry Bed

In this test case, the initial conditions are designed to generate two strong rarefaction waves travelling in opposite directions, producing a nearly dry bed at the centre of the modelled domain. The initial conditions are a water depth of $h_{\text{out}} = h_R = h = 1.0$ m throughout the channel and two opposite flow velocities of $u_L = -5.0$ ms^{-1} and $u_R = 5.0$ ms^{-1} to the left and the right, respectively, of the location of the discontinuity $x = 25$ m. Figure 3.9 shows free surface and flow velocity profiles at times $t = 0.0$ s and 2.5 s. Two rarefaction waves have developed propagating in opposite directions from the centre of the channel, with the depth dropping close to

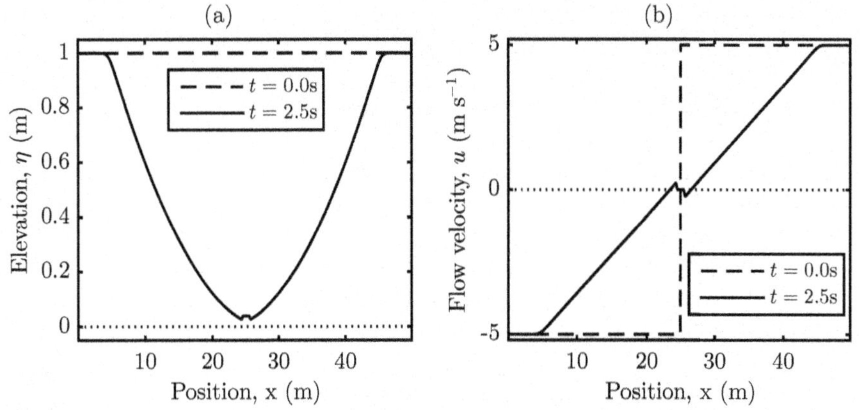

Fig. 3.9 Case 2: two rarefactions and nearly dry bed at initial conditions and $t = 2.5$ s. **a** Free surface elevation profiles; **b** flow velocity profiles

zero and velocity also close to zero at the centre itself. These profiles are remarkably similar to those obtained by Toro [15]. The present solver predicts a depth $h_* = 0.03907$ m and particle velocity $u_* = 0.0\ \mathrm{ms}^{-1}$ within the star region, in satisfactory agreement with the refined values of $h_* = 0.040564$ m and $u_* = 0.0\ \mathrm{ms}^{-1}$ given by Toro [15].

Case 3a and 3b - Right/Left Dry Bed Riemann Problem

The solver is run for a right-propagating dam break wave over a dry bed, with initial conditions such that the water depth to the left of the dam, located at $x = 20$ m, is $h_{\mathrm{out}} = 1.0$ m and 0.0 m to the right. Initially, the water is still. Unlike the wet bed case, a bore does not form; instead the free surface tends gradually towards the bed, in accordance with theory (see Toro [15]). The final profiles of free surface elevation and flow velocity in Fig. 3.10a, b are again almost identical to those obtained by Toro. Repeating the simulation, but with a left-facing dam break, gives mirror image results, shown in Fig. 3.10c, d, as would be expected.

Case 4 - Generation of Dry Bed

In this test, the left and right data states are chosen such that a dry bed is generated at the centre of the numerical domain, with two rarefaction waves travelling in opposite directions. The initial conditions are given by a water depth of $h = 0.1$ m throughout the domain and flow velocities of $u_L = -3.0\ \mathrm{ms}^{-1}$ and $u_R = 3.0\ \mathrm{ms}^{-1}$ to the left and right, respectively, of the discontinuity located at $x = 25$ m. Figure 3.11 presents the initial and final profiles along the channel of free surface elevation and flow velocity. The results are again in very satisfactory agreement with those of Toro [15], except in the 'dried out' region where Toro sets the depth and velocity to zero, whereas the present results are given in their 'raw' form.

Case 5 - Transcritical Flow over a Fixed Hump

This verification case tests the ability of the scheme to model the flow depth and velocity discontinuities that occur as free surface flow passes over a fixed bed hump, initially accelerating from subcritical (i.e. deep, slow) flow to become supercritical (shallow, fast) flow where the local Froude number is greater than unity, and then abruptly changing state back to subcritical flow by means of a hydraulic jump.

The bed profile, in metres, is given by

$$z_b(x) = \begin{cases} 0.2 - 0.05(x - 10)^2 & \text{if } 8 \leq x \leq 12, \\ 0.0 & \text{otherwise.} \end{cases} \tag{3.62}$$

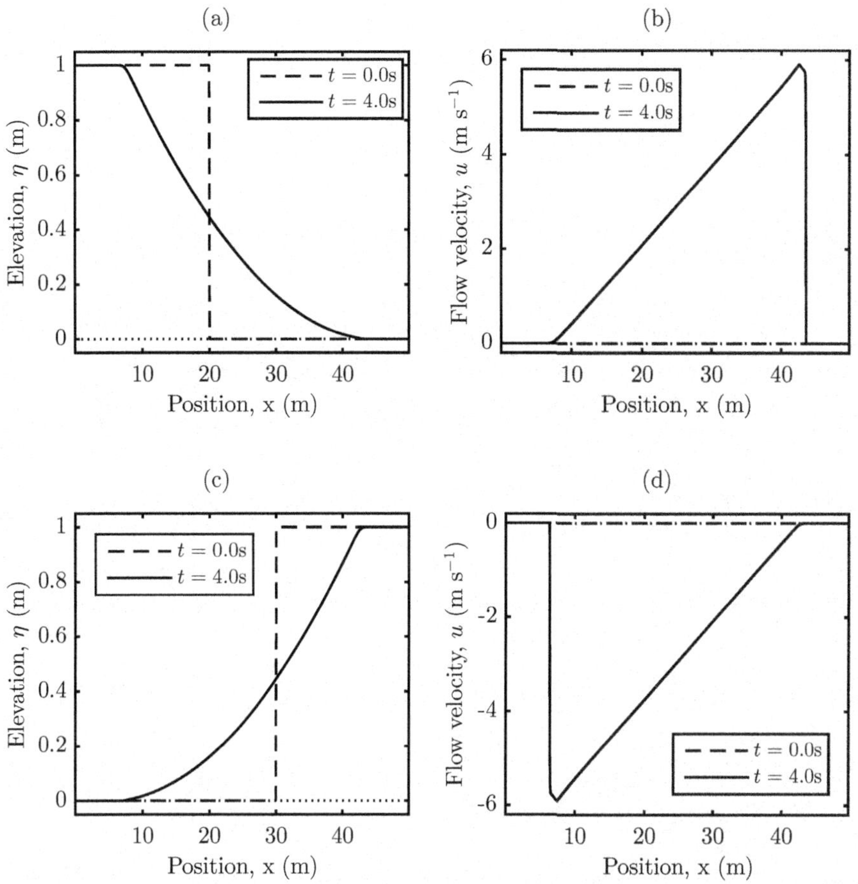

Fig. 3.10 Cases 3a and 3b: right and left dry bed Riemann problem at initial conditions and $t = 4.0\,$s. **a** and **c** free surface elevation profiles; **b** and **d** flow velocity profiles

The free surface elevation at the downstream boundary is set to $\eta = 0.33$ m and the flow rate is allowed to vary using a transmissive boundary condition. At the inlet, the discharge per unit width is prescribed as $q = uh = 0.18\ \text{m}^2\ s^{-1}$ and the depth is extrapolated. The bed is frictionless. Numerical parameters were set to $\Delta t = 0.01$ s and $\Delta x = 0.1$ m following convergence testing.

Figure 3.12 shows the bed elevation, free surface elevation, Froude number and flow rate profiles along the channel at steady state. We can see the hydraulic jump immediately downstream of the hump, where the Froude number becomes greater than unity. The results are in excellent agreement with the analytic solution provided by Goutal and Maurel [6], confirming that the present Godunov-type shallow flow solver is able to reproduce transcritical flows correctly.

Sampling of the results from the solver has a strong impact on the correct representation of the discharge at the jump itself. As explained by Ying and Wang [18],

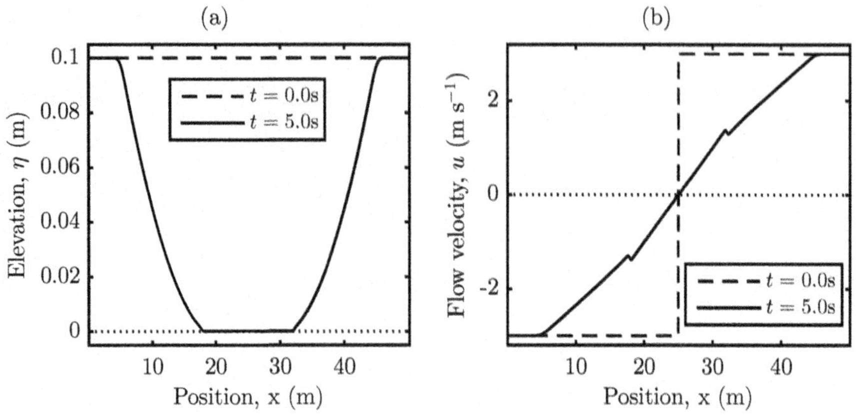

Fig. 3.11 Case 4: generation of a dry bed at initial conditions and $t = 5.0$ s. **a** Free surface elevation profiles; **b** flow velocity profiles

sampling the solution from the updated state vector **q** results in spurious values in the cells either side of the hydraulic jump. If, instead, the flux vector $\mathbf{f}(\mathbf{q})$ is sampled, the result agrees with the theoretical value of $q = 0.18$ m^2 s^{-1}. This is illustrated in Fig. 3.12c which shows the flow rate sampled from the data vector and Fig. 3.12d where the discharge is sampled from the flux vector.

3.4.8 Power from a Channel

The numerical solver is now used to simulate flow in a one-dimensional channel from which energy is extracted by a patch of enhanced roughness. The power P per unit width W dissipated by the patch, extending from $x = x_1$ to $x = x_2$, is given by integrating, over the length of the patch, the additional force experienced by the flow due to the presence of the turbines multiplied by the mass flow rate, i.e.

$$\frac{P}{W} = \int_{x_1}^{x_2} \rho(C_t)u^3 \,\mathrm{d}x, \tag{3.63}$$

where W is the width of the channel. This is implemented numerically as

$$\frac{P}{W} = \sum_{i=0}^{i=i_{\max}} \rho(C_i - C_d){u_i}^3 \Delta x \tag{3.64}$$

where C_i is the total bed roughness coefficient applied within the ith cell and C_d is the bed roughness coefficient of the natural channel.

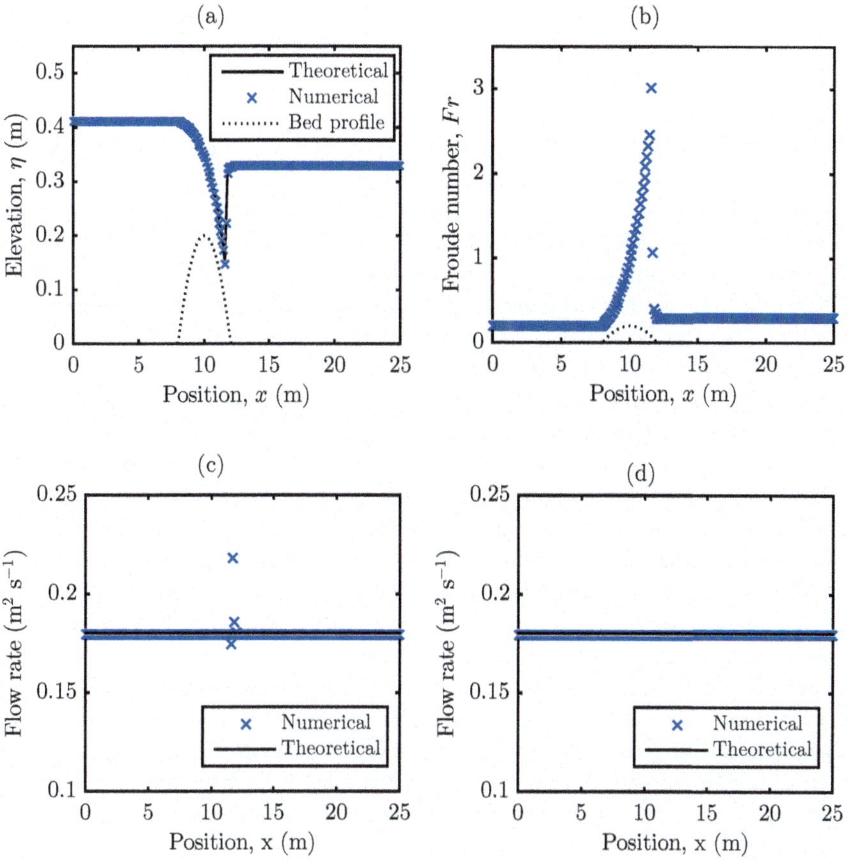

Fig. 3.12 Case 5: Hydraulic jump formation due to transcritical flow over a fixed bed hump. **a** Free surface elevation profile; **b** Froude number; **c** flow rate sampled from data state vector; **d** flow rate sampled from flux vector

3.4.9 Validation Test

The model is validated against a steady-state unidirectional flow example proposed by Bryden et al. [1], which is used to simulate depth and velocity profiles along an open channel in the absence of energy extraction from a patch. The channel has a flat, horizontal bed, and the flow is driven by a fixed head difference across the ends of the channel domain which is of length $L = 4000$ m. Boundary conditions are such that the flow depth is prescribed as 39.5297686733 m at the inlet (left-hand) end, and 39.2 m at the exit (right-hand) end. The value at the inlet was determined using the 40.0 m inlet value prescribed by Bryden et al. [1] and deducting the head loss due to entry conditions specified by Bryden et al. [1] which was calculated iteratively. Transmissive flux conditions are applied to the flow discharge at both ends. Initially, the flow discharge is set to 0.0 m^3s^{-1} per metre width throughout the computational domain, and the flow depth set to vary linearly over the length of the channel. The bed roughness is set using a Manning coefficient of $n = 0.035$ $sm^{-1/3}$, as used by Bryden et al. [1], using the relationship $C_d = gn^2/h^{1/3}$ for a one-dimensional channel (i.e. of infinite width such that the hydraulic radius $R = A/P_{er} \rightarrow h$ as the width $W \rightarrow \infty$, where P_{er} is the wetted perimeter). This gives a value of $C_d = 0.0035$ which is applied throughout the domain. Using a grid convergence test, it was found that a grid spacing of $\Delta x = 1$ m and a time step of $\Delta t = 0.02$ s gave converged results. Figure 3.13a shows the steady state depth and flow velocity profiles along the channel. There is an almost linear reduction in depth, and a corresponding monotonic increase in flow velocity in the downstream direction from end to end of the channel, as would be expected from basic hydrodynamics.

Next, a fence of turbines representing a patch of enhanced roughness was added in the middle of the channel, extending a total stream-wise distance of 100 m from $x_1 = 1950$ m to $x_2 = 2050$ m. It was assumed that the head difference across the ends of the channel was unchanged by the power extraction and the same prescribed depth boundary conditions were utilised as for the unexploited case. The additional roughness is expressed by raising the bed roughness coefficient by an additional coefficient C_t representing the presence of the turbines. The value of this additional coefficient is given by $C_t = \alpha C_d$, where α is a multiplier. Hence, the total roughness coefficient at the patch is $C_{\text{tot}} = C_d + C_t$. Figure 3.13b shows the steady state depth (solid lines) and velocity profiles (dashed lines) with the enhanced roughness present, for different values of α. In all cases, the effect of the local increase in roughness is to cause a step drop in depth (representing the local head driving a turbine, say) and a concomitant increase in flow velocity (accounting for conservation of mass in a one-dimensional channel, where the turbines are effectively acting as a fence across the channel).

Table 3.1 lists the power per unit channel width dissipated by the patch of enhanced roughness for natural bed roughness coefficient values of $C_d = 0.001$ and $C_d = 0.0035$, as calculated using (3.64). These values are in excellent agreement with power computations by Hammer [7] and Cao [2] using alternative finite difference solvers of the one-dimensional shallow water equations.

3.5 Parameter Study

A parameter study was carried out to examine the effect of uncertainty in natural bed roughness on the power extracted from a one-dimensional channel of length $L = 20$ km, depth $h_{\text{in}} = 50$ m and a flat, horizontal bed. The head drop along the channel was calculated assuming an inlet velocity of 3 ms^{-1} giving a head difference of $\Delta h_{channel} = -C_d u^2 L/(gh) = -2.75$ m for a bed roughness coefficient of $C_d = 0.025$ (i.e. the mean value assumed for C_d) and an average velocity $u = 5.2$ ms^{-1}. The upstream and downstream depths were prescribed as 51.375 m and 48.625 m respectively, to give an average depth (for an undisturbed channel) of 50 m, and

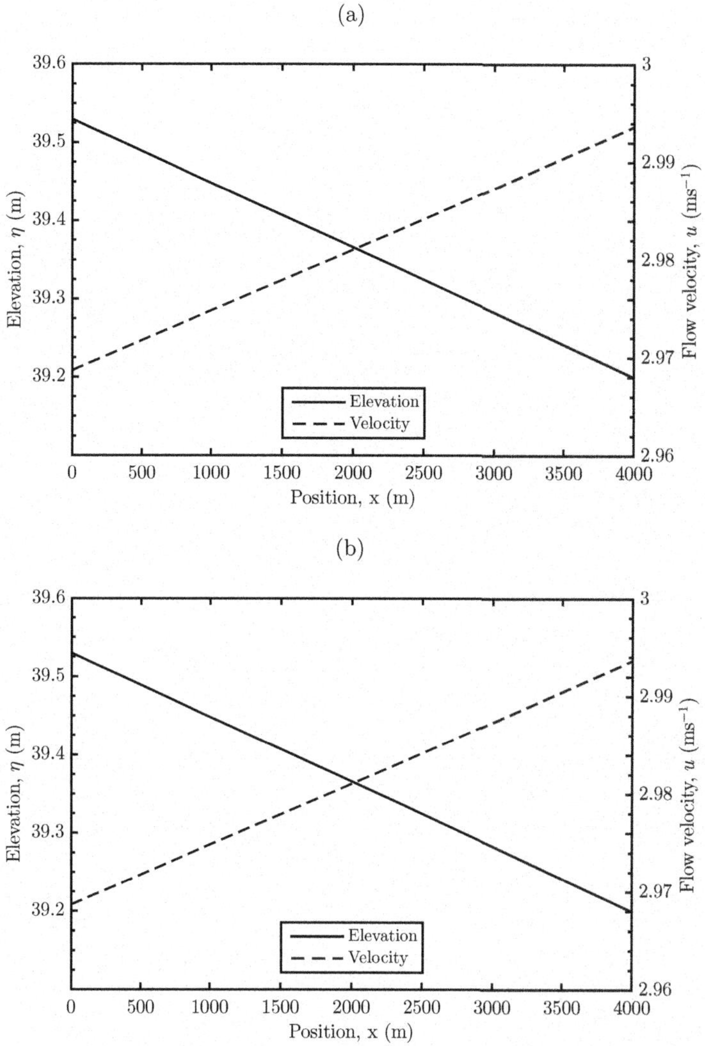

Fig. 3.13 Water depth and flow velocity profiles at steady state for a one-dimensional channel for: **a** no energy extraction; and **b** energy extraction over a patch of enhanced roughness in the center of the domain for a channel with natural bed roughness coefficient of $C_d = 0.0035$. The increase in roughness is given by $C_t = \alpha C_d$

transmissive flow conditions were again applied. Grid convergence tests were carried out, and the spatial increment and time step set to $\Delta x = 2.5$ m and $\Delta t = 0.05$ s, respectively, achieving a balance between accuracy of the solver and computational time.

For a given bed roughness coefficient C_d, the enhanced roughness of the patch C_t was varied and the instantaneous dissipated power calculated using (3.64) once

Table 3.1 Power dissipation per unit width calculated from the one-dimensional channel model for different bed roughness coefficients and different values of roughness multiplier α, compared with relevant values calculated by Hammer [7] and Cao [2]

Bed roughness coefficient, C_d	Multiplier α	Power dissipated per unit width (Wm^{-1})	Values by Hammer [7] ($C_d = 0.001$) and Cao [2] ($C_d = 0.0035$)
$C_d = 0.001$	65	2.6203×10^5	–
	70	2.6371×10^5	2.77562×10^5
	75	2.6483×10^5	–
	80	2.6547×10^5	2.81427×10^5
	85	2.6574×10^5	
$C_d = 0.0035$	70	1.4547×10^5	–
	75	1.4589×10^5	–
	80	1.4607×10^5	1.4696×10^5
	85	1.4605×10^5	–

steady state had been reached, taken to be at $t_{\text{end}} = 50,000$ s. The length of the patch was set to 1 km, with the patch located at the centre of the numerical domain. The model was run for an array of values of C_d and C_t covering the relevant solution space, and cubic spline interpolation used to refine the resulting look-up table of power values.

Figure 3.14a, b show the power surface and associated contour plot determined from the numerical model. For comparison, the equivalent analytic power surface and contour plot are plotted in Fig. 3.14c, d, using (3.12). The peak in the analytic model has been curtailed for easier comparison of the plots. The numerical and analytic plots agree well with one another, both qualitatively and quantitatively. Figure 3.14e, f show the difference in power calculated from the two different methods as a surface and a contour plot, respectively. There is very good agreement for almost all values of C_d and C_t, except for a small zone near the origin, as highlighted by the zoomed-in contour plot in Fig. 3.14f. It can be seen that discrepancies arise near the origin, in the absence of bed roughness $C_d = 0$ where the analytic model exhibits a singularity as $C_t \to 0$ (but has been cut off in Fig. 3.14c for ease of comparison to the numerical results). This phenomenon is not exhibited by the numerical model because it was run with non-zero values of natural bed roughness coefficient. The location of the optimal turbine drag C_t^*, shown as a dashed red line, is also predicted well by the analytic model.

The truncated normal PDF described by (3.13) was used as input distribution for the bed roughness coefficient, and the expected power has been calculated using (3.30) to 2nd order, as well as using the numerical PDF transfer method outlined in Sect. 3.3.2. Figure 3.15a, c and e show power dissipated as a function of turbine drag at a value of bed roughness coefficient of $C_d = 0.0025$, calculated from the numerical

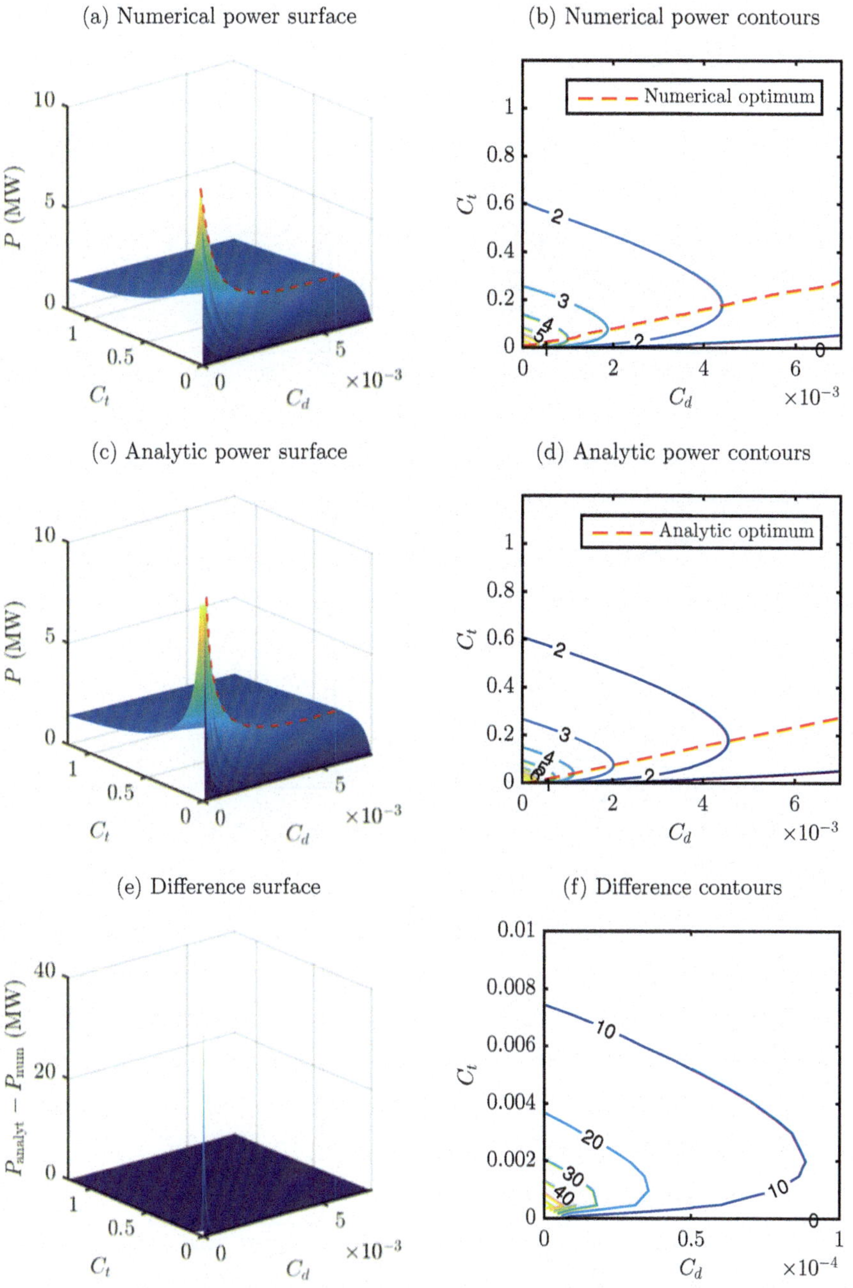

Fig. 3.14 Power (in MW) dissipated by enhanced bed roughness in a one-dimensional channel, forced by a static head difference as a function of natural bed roughness C_d and additional turbine drag C_t as a surface (**a**) and (**c**) and as contours (**b**) and (**d**) from the numerical and analytic models. The difference between the two, $P_{\mathrm{analyt}} - P_{\mathrm{num}}$ is shown in panels (**e**) and (**f**)

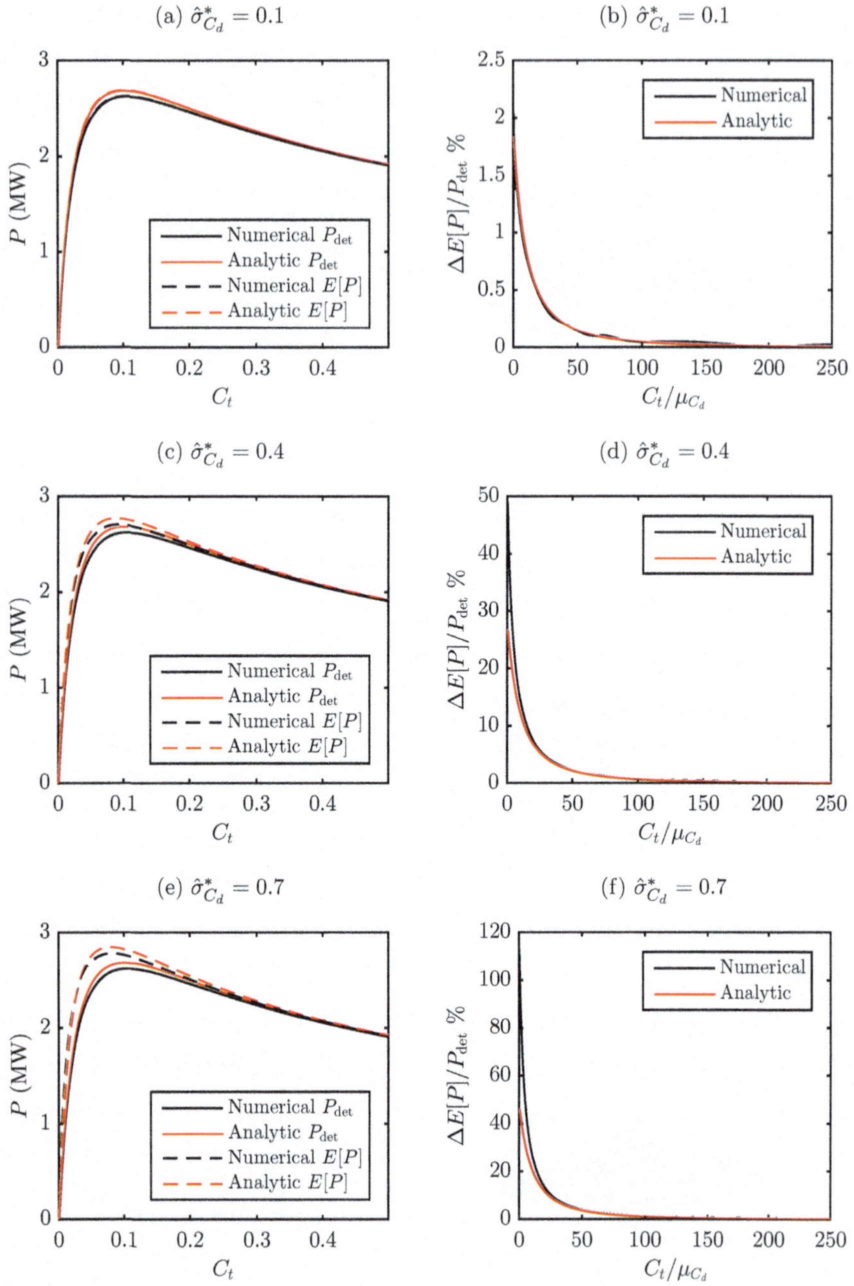

Fig. 3.15 Power and expected power (left-hand panels) and change in expected power as percentage of deterministic power P_{det} (right-hand panels) determined from the numerical simulations (black lines) and from the expansion method applied to the analytic model (red lines) for a mean bed roughness coefficient $\mu_{C_d} = 0.0025$ and $\hat{\sigma}_{C_d} = \{0.1, 0.4, 0.7\}$, i.e. $\hat{\sigma}^*_{C_d} = \{0.1, 0.382, 0.503\}$

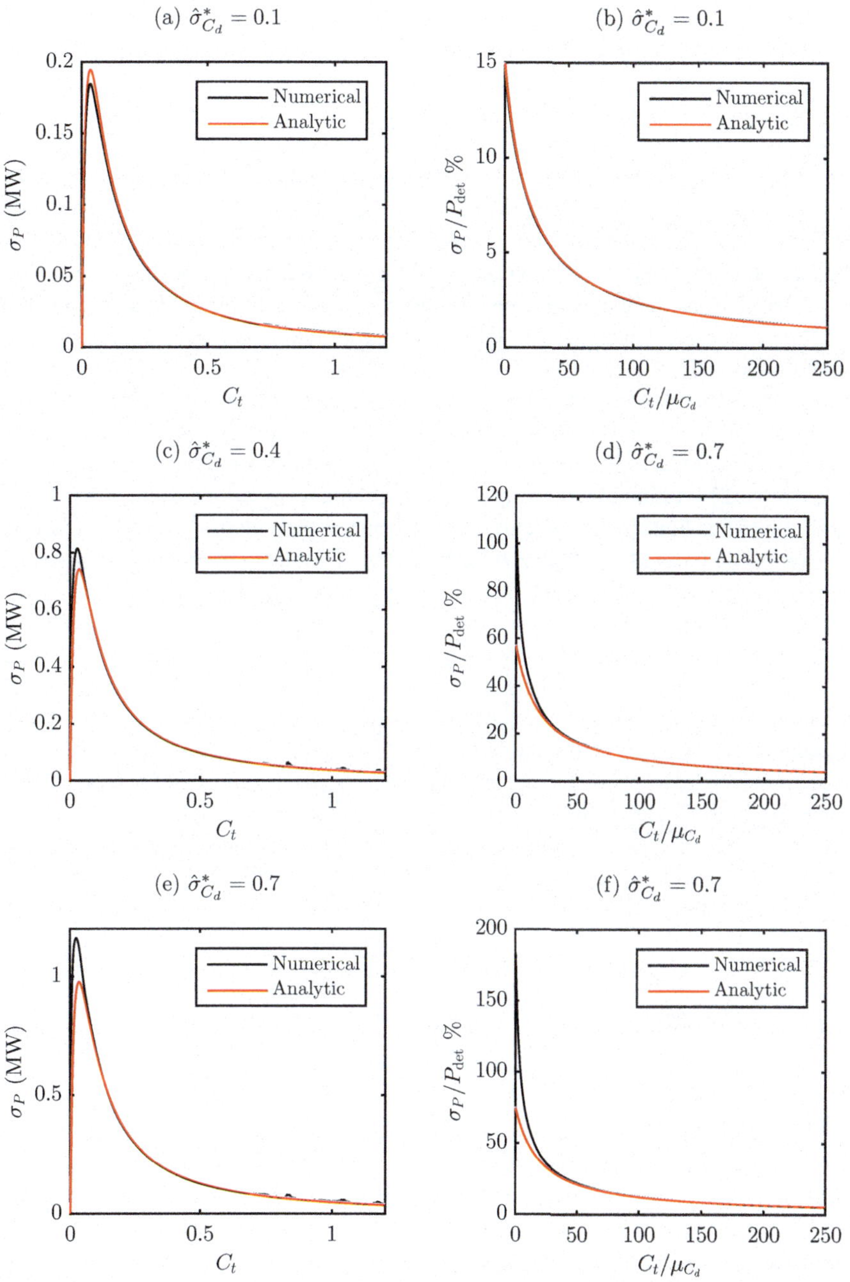

Fig. 3.16 Standard deviation (left-hand panels) and percentage standard deviation in power (right-hand panels) determined from the numerical simulations (black lines) and from the expansion method applied to the analytic model (red lines) for a mean bed roughness coefficient $\mu_{C_d} = 0.0025$ and $\hat{\sigma}_{C_d} = \{0.1, 0.4, 0.7\}$, i.e. $\hat{\sigma}^*_{C_d} = \{0.1, 0.382, 0.503\}$

model and the analytic model using (3.12). Alongside this, the expected power curves determined from the numerical PDF transfer method are plotted (dashed curves) for a mean bed friction coefficient $\mu_{C_d} = 0.0025$ and three values of relative standard deviation of the input "parent" normal distribution, $\hat{\sigma}_{C_d} = \sigma_{C_d}/\mu_{C_d} = \{0.1, 0.4, 0.7\}$ with associated truncated relative standard deviation values of $\hat{\sigma}^*_{C_d} = \{0.1, 0.382, 0.503\}$. The truncated standard deviation values were used in the expansion method to determine the expected power and also plotted in Fig. 3.15a, c and e (red dashed curves). It can be seen that uncertainty acts to increase the mean power, and that the value for bed roughness coefficient that maximises the power yield has shifted to a lower value of C_t. Slight discrepancies between the power values determined from the analytic and the numerical model can be observed, particularly for small values of C_t, as noted previously.

Figure 3.15b, d, and f show the change in expected power as percentage of deterministic power P_{det} for the same scenarios. The expansion technique again approximates well the effect of uncertainty on power though increasingly less so with increasing standard deviation owing to the truncation of the series after second order. In general, the expansion method slightly underestimates the relative change in expected power, particularly at low values of turbine drag, compared to the numerical results.

Figure 3.16a, c, and e show how the standard deviation in power changes with deployed turbine drag C_t for the same value of mean bed roughness coefficient. Figure 3.16b, d, and f show the relative standard deviation as a function of the scaled turbine drag. Again, a discrepancy between the expansion and the numerical PDF transfer methods becomes noticeable at low values of C_t, where the analytic model predicts a lower value of standard deviation than the numerical model. It is evident that higher-order terms in the expansion method become important for small C_t (and small μ_{C_d}), as is evident from the pronounced peak in the power surface (Fig. 3.14a). However, the expansion method is nonetheless able to capture trends in the impact of uncertainty in bed roughness coefficient on power, particularly for low values of input standard deviation.

3.6 Conclusions

This chapter has examined the effect of bed roughness uncertainty on power estimates for a hydro-kinetic turbine in steady, one-dimensional flow in a strait, modelled using the one-dimensional shallow water momentum equation. An analytic model was derived for the power dissipated by a patch of enhanced friction in a one-dimensional channel driven by a static head difference. This model was used to compare three methods for uncertainty propagation from an input probability density function in bed roughness coefficient to expected power and the standard deviation in power. These methods are an analytic PDF transfer, a numerical PDF transfer (which are shown to be largely equivalent), and an expansion method applied to the closed-form solution for power. A numerical model was developed, using a Godunov-type finite volume solver validated against standard benchmark tests. This model was used to compare the effectiveness of the expansion method to predict the leading-order effects of bed roughness uncertainty in a scenario where an approximate expression for the power in terms of bed roughness is available, as compared to power determined from a numerical model. The expansion method is found to be effective for approximating the effect of uncertainty, but is sensitive to ability of underlying power models to accurately capture the gradients of power with C_d.

The methods introduced in this chapter should be applicable to hydro-kinetic turbines placed in a steady stream, such as a natural river. Of the techniques considered the numerical model approach should prove useful to engineers tasked with power assessment in cases where the local bed conditions are poorly known. The next chapter extends the uncertainty modelling from turbines in steady flow to tidal stream turbines.

References

1. Bryden IG, Grinsted T, Melville GT (2004) Assessing the potential of a simple tidal channel to deliver useful energy. Appl Ocean Res 26(5):198–204
2. Cao B (2015) Tidal stream power from an array of devices. Bachelor's thesis, University of Edinburgh, United Kingdom
3. Fraccarollo L, Toro EF (1995) Experimental and numerical assessment of the shallow water model for two-dimensional dam-break type problems. J Hydraul Res 33(6):843–864
4. Garrett C, Cummins PF (2005) The power potential of tidal currents in channels. Proc Royal Soc A 461:2563–2572
5. Godunov SK (1959) Разностый метод численного расчета разрывных решений уравнений гидродинамики A difference method for numerical calculation of discontinuous solutions of the equations of hydrodynamics. Mat Sb 47(3):271–306
6. Goutal N, Maurel F (1997) Proceedings of the 2nd workshop on dam-break wave simulation. Technical Report HE-43/97/016/B, Electricité de France, Direction des études et recherches
7. Hammer F (2018) Tidal power resource assessment for a tidal strait. Master's thesis, University of Edinburgh, United Kingdom
8. Hirsch C (2007) Numerical computation of internal and external flows. Butterworth-Heinemann
9. Liang Q, Borthwick AGL (2009) Adaptive quadtree simulation of shallow flows with wet - dry fronts over complex topography. Comput Fluids 38(2):221–234
10. MATLAB (2018) version 9.3 (R2017b). The MathWorks Inc., Natick, Massachusetts
11. Papoulis A (1991) Probability, random variables, and stochastic processes. McGraw-Hill, New York
12. Rogers BD, Borthwick AGL, Taylor PH (2003) Mathematical balancing of flux gradient and source terms prior to using Roe's approximate Riemann solver. J Comput Phys 192(2):422–451
13. Rogers BD, Fujihara M, Borthwick AGL (2001) Adaptive Q-tree Godunov-type scheme for shallow water equations. Int J Numer Methods Fluids 35:247–280
14. Toro EF (1992) Riemann problems and the WAF method for solving the two-dimensional shallow water equations. Philos Trans Phys Sci Eng 338(1649):43–68
15. Toro EF (2001) Shock-capturing methods for free-surface shallow flows. Wiley
16. Toro EF (2009) Riemann solvers and numerical methods for fluid dynamics, 3rd edn. Springer, Berlin Heidelberg
17. Toro EF, Spruce M, Speares W (1994) Restoration of the contact surface in the HLL-Riemann solver. Shock Waves 4:25–34
18. Ying X, Wang S (2008) Improved implementation of the HLL approximate Riemann solver for one-dimensional open channel flows. J Hydraul Res 46(1):21–34

Chapter 4
Uncertainty in Background Friction

4.1 Introduction

In this chapter, the effect of uncertainty in the parameterisation of bed friction on estimates of extractable power is assessed in different analytic models for tidal energy extraction in which turbines are represented as either local or global enhanced bed roughness. Insight into the effect of the underlying physical assumptions on uncertainty propagation is developed by considering closed-form solutions for the power dissipated as predicted by three analytic models of tidal power assessment. The first model is that of Garrett and Cummins [5] (henceforth GC05), who derive an analytic solution for quasi-steady flow in a channel spanned completely by tidal turbines. Second, the impact of retaining inertia is explored by examining the solution to the same governing equation by Vennell [16] (henceforth V10). V10 is able to include inertia in a closed-form solution by making further approximations (see appendix of V10). Third and finally, the effect of flow diversion around the turbines is examined by considering [6] (henceforth GC13), who consider a circular turbine farm in laterally unconfined flow. Analytic solutions of these types have been shown to give predictions in satisfactory agreement with results from numerical models [8, 14]. Uncertainty is introduced to the value of background roughness coefficient C_d in these models and perturbation methods used to identify the leading-order effect of this uncertainty on the expected power dissipated by the turbines and the optimal channel design. Using the best estimate for the magnitude of the uncertainty in background roughness coefficient, quantitative estimates are provided of the effects of uncertainty on expected power dissipated and optimal channel design.

In Sect. 4.2, after a brief review of each model, uncertainty is introduced into the three theoretical models (GC05, V10 and GC13) and estimates obtained of the effect

---

An abridged version of this chapter has been published—see Kreitmair et al. [10].

M. J. Kreitmair, *The Effect of Uncertainty on Tidal Stream Energy Resource Estimates*, Springer Theses, https://doi.org/10.1007/978-3-030-57658-5_4

of uncertainty using perturbation methods. In Sect. 4.3, a best estimate is made of the relative standard deviation of the bed roughness coefficient (the ratio of the standard deviation in the value of C_d to its mean). Using this calibration and the leading-order solutions, the effects of uncertainty are investigated in Sect. 4.5 and compare the three models. Conclusions are drawn in Sect. 4.6.

4.2 Uncertainty in Theoretical Models

Fully-Spanned Channel (GC05 and V10)

In the model of GC05 (Fig. 4.1a) power is extracted from a channel of length L and depth h connecting two large bodies of water by means of a fence of turbines that fully spans the cross-section of the channel. The flow is driven in the simplest case by a sinusoidal tide producing a head difference between the ends of the channel, of amplitude a and angular frequency ω. Water is drawn in smoothly at speed u_0 at the entrance of the channel and exits as a jet at speed u_e. Furthermore, the channel is assumed sufficiently short compared to the tidal wavelength that the volume flux $Q = Au$, (where A is the cross-sectional area) is independent of distance along the channel x. These assumptions allow integration of the one-dimensional shallow water momentum equation (2.29) in non-conservative form, i.e.

$$\frac{\partial u}{\partial t} + u\frac{\partial u}{\partial x} + g\frac{\partial \zeta}{\partial x} = -\frac{C_d u^2}{h} - \frac{F_{\text{turb}}}{h}, \tag{4.1}$$

along the length of the channel to give, upon re-arranging and using $Q = Au$,

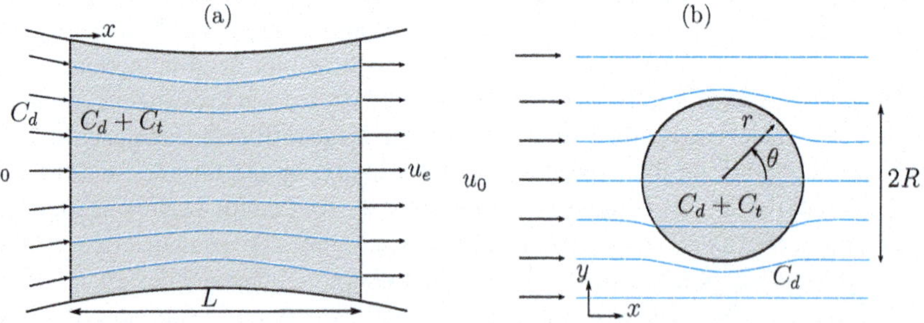

Fig. 4.1 Definition sketches for **a** the fully-spanned channel models of GC05 and V10 (adapted from [5]) and **b** the laterally unconfined model of GC13, where the shaded area is the region of increased bed friction ($C_d + C_t$, in which C_d is the background friction) representing the turbine farm of radius R with a uniform upstream velocity of u_0 in the x-direction. Streamlines are shown as blue dashed lines

$$\frac{\mathrm{d}Q}{\mathrm{d}t}\int_0^L \frac{1}{A}\,\mathrm{d}x + Q^2\int_0^L \frac{1}{2}\frac{\partial(1/A^2)}{\partial x}\,\mathrm{d}x + g\left(\zeta|_{x=L} - \zeta|_{x=0}\right) = \int_0^L \left[-\frac{C_d u^2}{h} - \frac{F_{\mathrm{turb}}}{h}\right]\mathrm{d}x. \tag{4.2}$$

This may be further rearranged to give

$$\gamma\frac{\mathrm{d}Q}{\mathrm{d}t} - ga\cos(\omega t) = -(\delta_0 + \delta_T)\,|Q|\,Q. \tag{4.3}$$

Here $\gamma = \int_0^L A^{-1}\,\mathrm{d}x$ is a geometric factor taking into account the varying cross-sectional area of the channel, t is time, the term $ga\cos(\omega t)$ is the driving pressure force due to the tide where g is gravity. On the right-hand side, $\delta_0 = \int_0^L C_d(hA^2)^{-1}\,\mathrm{d}x + (1/2){A_L}^{-2}$. The two terms comprising δ_0 account for the friction due to a given bed roughness coefficient C_d along the channel and the velocity head loss at the channel exit where the cross-sectional area is A_L. (The energy loss at the entrance is assumed to be negligible due to the slow drawing in of water at this end.) Energy dissipated due to power extraction is given by $\delta_T = \int_0^L C_t(hA^2)^{-1}\,\mathrm{d}x$, with turbines represented by a distributed roughness coefficient C_t. This coefficient represents the additional drag acting on the flow equivalent to the presence of the turbines.

By introducing the non-dimensional variables $t' = \omega t$, $Q' = Q\omega\gamma/(ga)$, $\lambda_0 = ga\delta_0/(\gamma\omega)^2$ and $\lambda_T = ga\delta_T/(\gamma\omega)^2$, GC05 obtain the expression

$$\frac{\mathrm{d}Q'}{\mathrm{d}t'} - \cos(t') = -(\lambda_0 + \lambda_T)\,|Q'|\,Q'. \tag{4.4}$$

The value of the parameter λ_0 determines the dynamic balance between natural friction and inertial forces acting in the channel. It represents the ratio of the combination of the natural drag losses and exit separation to acceleration, normalised by the driving amplitude [4]. Large values of λ_0 describe channels dominated by background friction and exit separation, i.e. shallow, short channels in which the flow may be considered to be quasi-steady. Small values of λ_0 correspond to channels in the inertial limit as would be the case for deep, long channels. The turbine drag parameter λ_T represents the equivalent drag force produced by the presence of turbines. Hence power dissipated by the turbines is given by multiplication of this drag term by the mass flow rate, i.e. $P = \rho\delta_T\,|Q|\,Q^2$, where ρ is the fluid density. The average power extracted by the turbines over a tidal cycle is then $\overline{P} = \rho\delta_T\overline{|Q|\,Q^2} = \rho(ga)^2(\gamma\omega)^{-1}\lambda_T\overline{|Q'|\,Q'^2}$, where the overline notation indicates time-averaging over the tidal period. The non-dimensional flow rate Q' is found by solving (4.4) and is, for a given head difference, a function of time and the total drag in the channel, i.e. $Q'(t', \lambda_0 + \lambda_T)$.

The Quasi-steady Limit (GC05)

GC05 derive an analytical solution for the average power in the quasi-steady state limit, i.e. for large values of λ_0. In this limit the acceleration term in (4.4) may be neglected and the non-dimensional volumetric flux may then be approximated by $\left|Q'\right| = (\lambda_0 + \lambda_T)^{-1/2} \left|\cos t'\right|^{1/2}$. The corresponding average power produced by the turbines becomes

$$P_{\mathrm{GC05}} = P_0 \frac{\lambda_T}{(\lambda_0 + \lambda_T)^{3/2}}, \tag{4.5}$$

where $P_0 = \beta_2 \rho (ga)^2/(\gamma\omega)$ is the dimensional multiplier for the power and $\beta_2 = \overline{|\cos t'|^{3/2}} \approx 0.56$ accounts for time-varying head difference and the subscript 2 denotes quadratic friction.

To introduce uncertainty in background friction, λ_0 is expressed as a random variable with an expected value of μ_{λ_0} and random, zero-mean fluctuation $\Delta\lambda_0$ about this value, such that $\lambda_0 = \mu_{\lambda 0} + \Delta\lambda_0$. Provided the fluctuation is small compared to the mean, the power produced by the turbines may be expressed in terms of λ_0 by expanding (4.5) as a Taylor series in $\Delta\lambda_0$ about the deterministic case ($\Delta\lambda_0 = 0$) as

$$\begin{aligned}\frac{1}{P_0} P_{\mathrm{GC05}} =& \frac{\lambda_T}{(\mu_{\lambda_0} + \lambda_T)^{3/2}} - \frac{3}{2} \frac{\lambda_T}{(\mu_{\lambda_0} + \lambda_T)^{5/2}} \Delta\lambda_0 + \frac{15}{8} \frac{\lambda_T}{(\mu_{\lambda_0} + \lambda_T)^{7/2}} \Delta{\lambda_0}^2 \\ &- \frac{35}{16} \frac{\lambda_T}{(\mu_{\lambda_0} + \lambda_T)^{9/2}} \Delta{\lambda_0}^3 + \frac{315}{128} \frac{\lambda_T}{(\mu_{\lambda_0} + \lambda_T)^{11/2}} \Delta{\lambda_0}^4 + \mathcal{O}\left(\Delta{\lambda_0}^5\right).\end{aligned} \tag{4.6}$$

Higher-order terms are included in the series, which converges for sufficiently small $\Delta\lambda_0$.

Expected Power Applying the expectation operator, the expected power extracted, correct to fourth order in $\Delta\lambda_0$, is given by

$$\begin{aligned}\frac{1}{P_0} \mathrm{E}[P_{\mathrm{GC05}}] =& \frac{\lambda_T}{(\mu_{\lambda_0} + \lambda_T)^{3/2}} + \frac{15}{8} \frac{\lambda_T}{(\mu_{\lambda_0} + \lambda_T)^{7/2}} \sigma_{\lambda_0}^2 - \frac{35}{16} \frac{\lambda_T}{(\mu_{\lambda_0} + \lambda_T)^{9/2}} \mu_{\lambda_0,3} \\ &+ \frac{315}{128} \frac{\lambda_T}{(\mu_{\lambda_0} + \lambda_T)^{11/2}} \mu_{\lambda_0,4} + \mathcal{O}\left(\mathrm{E}[\Delta{\lambda_0}^5]\right),\end{aligned} \tag{4.7}$$

where ${\sigma_{\lambda_0}}^2 = \mathrm{E}[\Delta{\lambda_0}^2]$ is the variance in background friction parameter, $\mu_{\lambda_0,3} = \mathrm{E}[\Delta{\lambda_0}^3]$ the skewness, and $\mu_{\lambda_0,4} = \mathrm{E}[\Delta{\lambda_0}^4]$, is the kurtosis. The second term of the series (4.6) vanishes as the random fluctuation $\Delta\lambda_0$ is symmetric about the mean. The first term in the expansion is simply the deterministic power removed by the turbines in a channel (4.5) at the mean background friction parameter μ_{λ_0}. The second term is a stochastic correction to the power resulting from considering a distribution of λ_0 values that are spread about the mean μ_{λ_0} with a standard deviation of σ_{λ_0}.

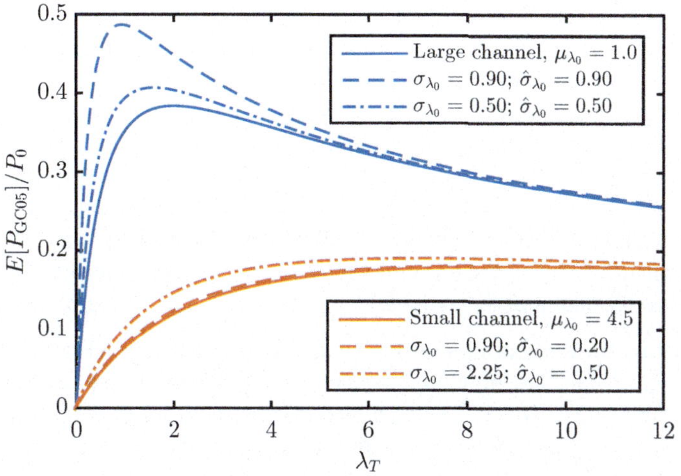

Fig. 4.2 Expected power produced by turbines in two fully-spanned tidal channels (GC05) with mean background friction parameter values of $\mu_{\lambda_0} = 1.0$ (representative of a large and deep channel) and $\mu_{\lambda_0} = 4.5$ (representative of a small channel). Power from equivalent deterministic channels is shown as solid lines. The dashed lines show the expected power from the two channels at the same value of standard deviation in background friction parameter $\sigma_{\lambda_0} = 0.90$. The dot-dashed lines have the same value of relative standard deviation $\hat{\sigma}_{\lambda_0} = \sigma_{\lambda_0}/\mu_{\lambda_0} = 0.50$

Higher-order terms take account of corrections due to further moments of the input probability density function, such as skewness and kurtosis.

The expected power (4.7) truncated at second order is shown as a function of turbine drag parameter λ_T in Fig. 4.2 for two channels with different values of the mean background friction parameter μ_{λ_0}. The first channel, with $\mu_{\lambda_0} = 1.0$, corresponds to a large and deep channel, and the second, with $\mu_{\lambda_0} = 4.5$, to a small channel with a high flow velocity [17]. It is clear that, regardless of the mean channel drag parameter or the value of turbine drag, uncertainty in λ_0 acts to increase expected power (dashed and dot-dashed lines) from that calculated using the deterministic model (continuous lines) and more so for greater σ_{λ_0} values. This effect is greatest for $\lambda_T = 2\mu_{\lambda_0}/5$, which maximises the second term in (4.7), but remains positive for all values of λ_T, reducing in strength as λ_T increases (and the effect of background roughness becomes less important).

This increase in expected power is a result of the inverse relationship between power (4.5) and bed friction parameter λ_0. Neglecting the inertial term in (4.4) (by assumption of the quasi-steady limit) requires that the head difference driving the flow is balanced solely by dissipation due to the total channel drag $\lambda_{\text{tot}} = \lambda_0 + \lambda_T$. Hence the flow rate Q is inversely proportional to λ_{tot} and, for a given driving head, Q must grow increasingly fast as total channel drag reduces, i.e. $\partial^2 Q/\partial\lambda_{\text{tot}}^2 > 0$. Consequently, a small reduction in bed roughness parameter away from the mean $\Delta\lambda_0^- < 0$ results in dissipation of a greater amount of power by the turbines, $\Delta P_{\text{GC05}}^- > 0$. Similarly, a small increase in bed roughness parameter of the same magnitude $\Delta\lambda_0^+ > 0$

yields reduction by an amount of power $\Delta P^{+}_{\mathrm{GC05}} < 0$ smaller than before, i.e. $|\Delta P^{+}_{\mathrm{GC05}}| < \Delta P^{-}_{\mathrm{GC05}}$. Assuming a symmetric probability density function for λ_0, the expected power, given by $\mathrm{E}[P_{\mathrm{GC05}}] = P_{\mathrm{GC05}}(\lambda_0 = \mu_{\lambda_0}) + (\Delta P^{-}_{\mathrm{GC05}} + \Delta P^{+}_{\mathrm{GC05}})/2$, is then necessarily greater than the deterministic power. In other words, due to the dynamic balance between driving head and channel drag, the power curve has a positive second derivative with respect to the channel bed roughness λ_0. This convexity results in an asymmetric power dissipation for symmetric perturbations in λ_0 and thus an increase in the expected power (cf. Jensen's inequality, which states that a convex transformation of the mean of a random variable is less than the mean of the convex transformation of the variable).

Optimal Turbine Drag In addition to a change in expected power, Fig. 4.2 also shows a shift in optimal turbine drag due to uncertainty. In the absence of uncertainty, this optimum occurs at a value of turbine drag that is twice the mean background friction parameter, $\lambda_{T}{}^{*}_{\mathrm{det}} = 2\mu_{\lambda 0}$. However, with increasing σ_{λ_0} the maximum shifts to lower values of λ_T. An analytical expression for the optimal turbine drag $\lambda_{T}{}^{*}_{\mathrm{stoch}}$ may be found by maximising (4.7) with respect to λ_T such that, to first order,

$$\lambda_{T}{}^{*}_{\mathrm{stoch}} = 2\mu_{\lambda 0}\left[1 - \frac{5}{6}\hat{\sigma}^2_{\lambda}\right] + \mathcal{O}(\mathrm{E}[\Delta\lambda_0{}^3]), \tag{4.8}$$

where $\hat{\sigma}_{\lambda_0} = \sigma_{\lambda_0}/\mu_{\lambda_0}$ is the relative standard deviation. The optimal turbine drag reduces linearly with the variance of the bed friction parameter. This may be understood by perturbing around the deterministic optimum, so that $\lambda_T{}^{*} = \lambda_{T}{}^{*}_{\mathrm{det}} + \Delta\lambda^{*}_{T}$. Along the optimum, $\partial\mathrm{E}\left[P_{\mathrm{GC05}}\right]/\partial\lambda_T = 0$. Expanding this identity around $\lambda_{T}{}^{*}_{\mathrm{det}}$ gives, after some manipulation,

$$\Delta\lambda^{*}_{T} = -\frac{1}{2}\frac{P_{\lambda_0\lambda_0\lambda_T}\left(\mu_{\lambda 0}, \lambda_{T}{}^{*}_{\mathrm{det}}\right)}{P_{\lambda_T\lambda_T}\left(\mu_{\lambda 0}, \lambda_{T}{}^{*}_{\mathrm{det}}\right)}\sigma^2_{\lambda}, \tag{4.9}$$

where P corresponds to P_{GC05} and the subscripts denote differentiation. The change in optimal turbine drag $\Delta\lambda^{*}_{T}$ depends on the sign of $P_{\lambda_0\lambda_0\lambda_T}$ (the change in the convexity of the power curve with turbine drag) and the sign of $P_{\lambda_T\lambda_T}$ (the convexity of the power with respect to turbine drag) calculated at the deterministic optimum to leading-order of approximation. Here $P_{\lambda_T\lambda_T} < 0$ because of the maximum. The effect of reducing λ_T is to lower the total channel drag, making the flow rate and hence the power more sensitive to the bed friction parameter. At lower values of λ_{tot}, the increase in power becomes relatively larger than the decrease in power for a fluctuation $\Delta\lambda_0$, and the change in the expected power increases ($\partial^3 P/\partial\lambda^3_{\mathrm{tot}} < 0$). It is therefore optimal in the presence of background friction uncertainty to choose a lower value of λ_T in order to harness better the uncertain power.

Uncertainty in Power The variance in power, $\sigma_P{}^2 = \mathrm{E}\left[(P - \mathrm{E}[P])^2\right]$, may be evaluated using (4.5) to give, to leading-order

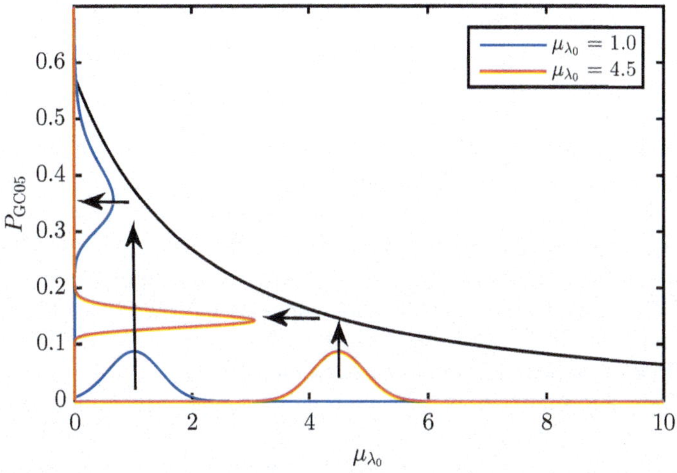

Fig. 4.3 Mapping of uncertainty from background friction λ_0 to power P_{GC05} via the transfer function of the power curve for a fully-spanned tidal channel (GC05). For both values of the mean background friction $\mu_{\lambda_0} = 1.0$ (representative of a large and deep channel) and $\mu_{\lambda_0} = 4.5$ (representative of a small channel) the standard deviation in λ_0 is $\sigma_{\lambda_0} = 0.45$ and the turbine drag is $\lambda_T = 3.0$

$$\sigma_P{}^2 = \frac{9}{4} P_0^2 \frac{\lambda_T{}^2}{(\mu_{\lambda 0} + \lambda_T)^5} \sigma_{\lambda 0}{}^2 + \mathcal{O}(\mathrm{E}[\Delta\lambda_0{}^3]) \tag{4.10}$$

$$= \left(P_{GC05}(\lambda_0 = \mu_{\lambda_0})\right)^2 \frac{9}{4} \frac{\hat{\sigma}^2_{\lambda_0}}{(1 + \lambda_T/\mu_{\lambda_0})^2} + \mathcal{O}(\mathrm{E}[\Delta\lambda_0{}^3]). \tag{4.11}$$

The greater the total mean drag in the channel $\mu_{\lambda_0} + \lambda_T$, the smaller the standard deviation in power. This may be understood by considering the mapping of the probability density function of the background friction parameter f_{λ_0} to the probability density function of power f_P: $f_P(P) = f_{\lambda_0}(\lambda_0(P))/|dP/d\lambda_0|$. Figure 4.3 illustrates this mapping. Probability density functions are shown for two different values of mean background friction, $\mu_{\lambda_0} = 1.0$ and $\mu_{\lambda_0} = 5.0$, for the same standard deviation of $\sigma_{\lambda_0} = 0.45$ and at a turbine drag of $\lambda_T = 3.0$. The greater the value of μ_{λ_0}, the smaller the standard deviation in power, due to the smaller gradient in the transfer function. Figure 4.3 shows that, despite symmetric input probability density functions for λ_0, the corresponding probability density functions for the power values are asymmetric. Propagation through the nonlinear transfer function has generated (positive) skewness in power. It is worth further noting that the probability distribution of total channel friction is technically not allowed to have zero negative values. Singular or complex values of power implied by zero or negative values of total channel friction are automatically avoided because the analysis is restricted to leading-order terms in uncertainty.

The Effect of Inertia (V10)

The quasi-steady limit in the previous section applies to channels in which friction dominates inertia in the dynamic balance of the channel, i.e. in the limit of large λ_0 values, and the inertial term in (4.4) may be neglected. Relaxation of the quasi-steady assumption leads to a different behaviour of the power potential of the channel under bed roughness uncertainty. The effect of retaining inertia in the channel dynamics is explored by considering the solution presented in the appendix of Vennell [16] (V10). Therein an analytic solution is derived to an approximation of (4.4) which retains the inertial term. Furthermore, the quadratic drag term is replaced with a linear drag term which ensures the same average power is dissipated by the turbines over a tidal cycle, a process known as Lorentz linearisation [15, 19].

Following this approach, and assuming a sinusoidal driving tide of single frequency ω as before, the drag term $(\lambda_0 + \lambda_T)\left|Q'\right| Q'$ in (4.4), where Q' is the non-dimensional flow rate, may be replaced with KQ' such that $(\lambda_0 + \lambda_T)\overline{|Q'|\, Q'^2} = K\overline{Q'^2}$, where $Q' = Q'_0 \sin(t' - \phi_Q)$ and ϕ_Q is the phase lag of the flow rate to the driving head difference between the ends of the channel. The coefficient K may be evaluated as $K = 8(\lambda_0 + \lambda_T)Q'_0/(3\pi)$. The resulting linearised governing equation gives (V10)

$$Q'_0 \cos(t' - \phi_Q) - \cos(t') \approx \frac{8}{3\pi}(\lambda_0 + \lambda_T)Q_0'^{\,2} \sin(t' - \phi_Q), \tag{4.12}$$

and may be solved to give the solutions (V10)

$$Q'_0 = \frac{\left(\sqrt{4\lambda_{\text{eq}}^2 + 1} - 1\right)^{1/2}}{\sqrt{2}\lambda_{\text{eq}}} \quad \text{and} \quad \phi_Q = \tan^{-1}\left(\frac{1}{\lambda_{\text{eq}} Q'_0}\right), \tag{4.13}$$

where the equivalent total friction parameter $\lambda_{\text{eq}} \equiv 8(\lambda_0 + \lambda_T)/(3\pi)$. Finally, the power produced by the turbines for this model is given by

$$P_{\text{V10}} = \frac{P_0}{\beta_2}\lambda_T \frac{8}{3\pi} Q_0'^{\,3}\, \overline{\sin(t' + \phi_Q)^2} = \frac{P_0}{\beta_2}\frac{4}{3\pi}\lambda_T \frac{\left(\sqrt{4\lambda_{\text{eq}}^2 + 1} - 1\right)^{3/2}}{\left(\sqrt{2}\lambda_{\text{eq}}\right)^3}. \tag{4.14}$$

As before, the magnitude of λ_0 defines the dynamic balance in the channel: small values indicate a channel that is dominated by inertia, whereas large values of λ_0 imply that background friction dominates. In the limit $\lambda_0 \to \infty$ (the quasi-steady limit) the same result as obtained by GC05 (4.5) is recovered, and in the limit of $\lambda_0 \to 0$ (the inertial limit) the following expression is obtained for the power produced by the turbines

$$P_{V10} \to \frac{P_0}{\beta_2} \frac{\sqrt{3\pi}}{256\sqrt{2}} \frac{\left(\sqrt{9\pi^2 + 256\lambda_T^2} - 3\pi\right)^{3/2}}{\lambda_T^2} \quad \text{as } \lambda_0 \to 0, \tag{4.15}$$

which is independent of λ_0.

Expected Power The expected power for V10 is determined in the same way as that for GC05. Power in V10 (4.14) is expanded as a Taylor series in $\Delta\lambda_0$ about the deterministic case $\lambda_0 = \mu_{\lambda_0}$, truncated to second order. Shorthand $\lambda_{\text{eq}} \equiv 8(\lambda_0 + \lambda_T)/(3\pi)$ and $\tilde{\mu} \equiv \sqrt{4\mu_{\lambda_{\text{eq}}}^2 + 1}$ are used to reduce clutter. The Taylor expansion is performed in terms of λ_{eq}, i.e. $\lambda_{\text{eq}} = \mu_{\lambda_{\text{eq}}} + \Delta\lambda_{\text{eq}}$ where $\mu_{\lambda_{\text{eq}}} = 8(\mu_{\lambda_0} + \lambda_T)/(3\pi)$ and $\Delta\lambda_{\text{eq}} = 8\Delta\lambda_0/(3\pi)$. Noting that the variance of the equivalent channel drag is $\sigma_{\lambda_{\text{eq}}}^2 = (8\sigma_{\lambda_0}/(3\pi))^2$ due to the linear transformation, the expected power is given by

$$\begin{aligned}\frac{\text{E}[P_{V10}]}{P_0} =& \frac{4}{3\pi}\lambda_T \frac{(\tilde{\mu}-1)^{3/2}}{(\sqrt{2}\mu_{\lambda_{\text{eq}}})^3} \\ &+ \frac{4}{3\pi}\lambda_T \frac{3\left(10\mu_{\lambda_{\text{eq}}}^4(\tilde{\mu}-4) + \mu_{\lambda_{\text{eq}}}^2(19\tilde{\mu}-27) + 4(\tilde{\mu}-1)\right)}{2\sqrt{2}\mu_{\lambda_{\text{eq}}}^5\left(4\mu_{\lambda_{\text{eq}}}^2+1\right)^{3/2}\sqrt{\tilde{\mu}-1}}\sigma_{\lambda_{\text{eq}}}^2 \\ &+ \mathcal{O}(\text{E}[\Delta\lambda_{\text{eq}}{}^3]).\end{aligned} \tag{4.16}$$

The first term is simply the power calculated from the V10 model (4.14) at a drag of $\lambda_{\text{eq}} = \mu_{\lambda_{\text{eq}}}$. The second term indicates the leading-order response of the model to uncertainty. The change in expected power as a fraction of deterministic power $(\text{E}[P_{V10}] - P_{\text{det}})/P_{\text{det}}$ changes sign from negative to positive at a value of $\mu_{\lambda_{\text{eq}}} = 0.420$, i.e. $\mu_{\lambda_0} + \lambda_T = 0.495$.

Figure 4.4a shows the change in the expected power, correct to second order in σ_{λ_0}, per unit relative variance $\sigma_{\lambda_0}^2$ (also known as the coefficient of variation) in λ_0 as a function of the turbine drag scaled by the mean background friction parameter. For increasing μ_{λ_0}, the effect of background friction uncertainty approaches that of GC05, as illustrated by the different colour curves. Furthermore, as the value of mean background friction μ_{λ_0} is reduced, the change in expected power drops to zero (see the line for $\mu_{\lambda_0} = 0.1$), reflecting the independence of power from background friction in the limit of small λ_0. In short, Fig. 4.4 indicates that inertia reduces the effect of uncertainty on expected power. The transition from the quasi-steady to the inertial limit can be non-monotonic. For channels with background friction $\mu_{\lambda_0} > \lambda_{\text{transition}}$ with $\lambda_{\text{transition}} = 0.495$, the change in expected power is positive for all values of λ_T and the flow dynamics are dominated by the effect of the total channel drag $\lambda_{\text{tot}} = \lambda_0 + \lambda_T$. For values of μ_{λ_0} below $\lambda_{\text{transition}}$, the change in expected power becomes negative for values of turbine friction given by $\lambda_T < \lambda_{\text{transition}} - \mu_{\lambda_0}$, as may be seen from the curves with $\mu_{\lambda_0} = 0.1$ and 0.3. This behaviour may be understood by considering the flow rate Q' as a function of bed friction parameter for an undisturbed

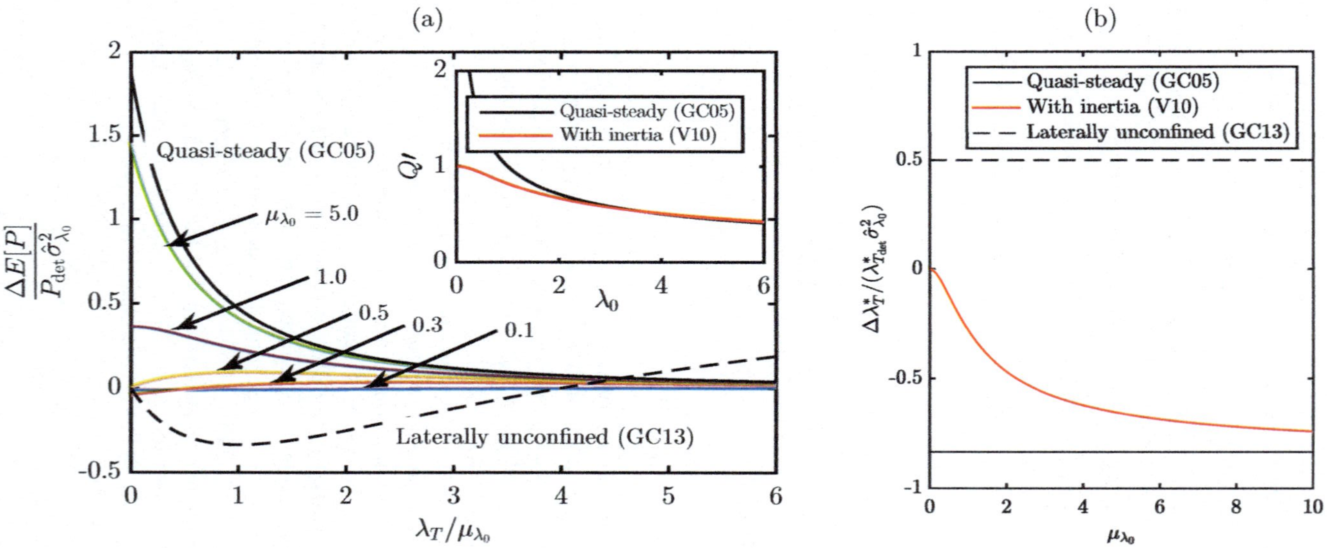

Fig. 4.4 Relative change in expected power per unit relative variance in background friction $\hat{\sigma}^2_{\lambda_0}$ as a function of turbine drag scaled with mean background friction at different values of μ_{λ_0} (**a**) and relative change in optimal turbine friction in the presence of uncertainty in the background friction parameter (**b**) for a fully-spanned tidal channel (GC05 and V10). In panel a, for the GC05 model (black line) the lines from Fig. 4.2 reduce to the same curve upon scaling. The inset shows the non-dimensional flow rate $Q' = \omega c/(ga)Q$ as a function of channel friction parameter λ_0 for an undisturbed channel, i.e. $\lambda_T = 0$

channel, shown in the inset in Fig. 4.4a. The singular limit as $\lambda_0 \to 0$ in GC05, is avoided by inertia in V10. Due to the bounded nature of the flow rate at low values of channel friction, a small decrease in bed roughness $\Delta\lambda_0^-$ will only slightly increase the flow rate because the channel is inertia-dominated ($\Delta P > 0$ but small). On the other hand, an increase $\Delta\lambda_0^+$ will be less affected by inertia, as the move is towards the drag-dominated regime ($\Delta P < 0$ and high). The expected power is therefore lower than in the deterministic case for $\mu_{\lambda_0} < \lambda_{\text{transition}} = 0.495$, where $\lambda_{\text{transition}}$ demarcates the transition between inertia-dominated and drag-dominated channels.

Optimal Turbine Drag In the V10 model, as for the GC05 model, the optimal turbine tuning changes upon introduction of uncertainty in λ_0. The turbine drag which maximises the expected power (4.16) was found numerically by using a Newton-Raphson algorithm to find the value of λ_T (in the limit of a small standard deviation in λ_0, to be consistent with the expansions in the other sections) which satisfies $\partial(4.16)/\partial\lambda_T = 0$ and gave a negative second derivative. Figure 4.4b shows the relative change in the optimal turbine drag λ_T^* per unit variance in λ_0, as a function of the mean bed friction coefficient. In the limit of zero background friction $\mu_{\lambda_0} \to 0$, the optimal turbine drag is unaffected because the flow behaviour is dominated by inertia. At very large values of the mean bed friction coefficient, the system becomes dominated by friction and the V10 model asymptotically approaches the quasi-steady limit of GC05.

Uncertainty in Power The variance for V10, i.e. $\sigma_{P_{V10}}^2 = \mathrm{E}[P_{V10}^2] - \mathrm{E}[P_{V10}]^2$, is given by

$$\sigma_{P_{V10}}^2 = \lambda_T^2 \frac{8\left(1 - \tilde{\mu} + (7 - 5\tilde{\mu})\mu_{\lambda_{\text{eq}}}^2 + (13 - 5\tilde{\mu})\mu_{\lambda_{\text{eq}}}^4 + 4\mu_{\lambda_{\text{eq}}}^6\right)}{\pi^2 \tilde{\mu}^3 \mu_{\lambda_{\text{eq}}}^8} + \mathcal{O}(\mathrm{E}[\Delta\lambda_{\text{eq}}{}^3]). \tag{4.17}$$

Compared to the quasi-steady limit (GC05), in which the relative standard deviation in power as a fraction of the standard deviation in background friction is a monotonically decreasing function of turbine drag scaled with mean background friction (cf. (4.11)), inertia reduces the effect of uncertainty, as illustrated in Fig. 4.5. As the dependence of power on background friction is reduced in the inertia-dominated regime, the resulting variance in power reduces also.

Laterally Unconfined Turbine Farm (GC13)

For turbine farms that do not span the channel completely, not all of the flow in the channel passes through the turbines, instead part of it is diverted around the turbines as bypass flow. In such cases, bed friction acts not only to reduce the flow speed in the channel, but also to funnel the flow through the turbine farm by resisting the bypass flow. These competing effects are explored by considering the model of Garrett and

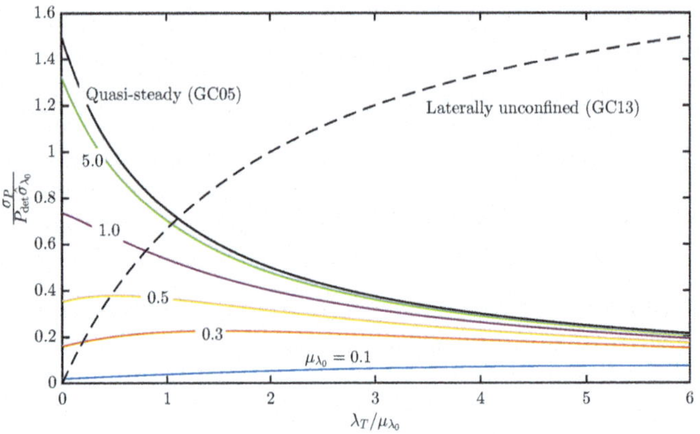

Fig. 4.5 Relative standard deviation in power per unit relative standard deviation in background friction $\hat{\sigma}_{\lambda_0}$ as a function of the turbine drag scaled with mean background friction. For the model retaining inertia (V10), this is a function of mean background friction and has been plotted at different values of μ_{λ_0}

Cummins [6] (GC13). In this model, energy extraction by a tidal farm is represented by a localised increase in bed roughness within a circular area of radius R in a steady flow of far-field current of u_0 in the x-direction and no lateral confinement by channel walls or similar (see Fig. 4.1b). The dynamic balance of the system is described by the shallow-water equations,

$$\frac{\partial \mathbf{u}}{\partial t} + \mathbf{f} \times \mathbf{u} + \mathbf{u} \cdot \nabla \mathbf{u} + g\nabla\zeta = -\frac{C_d}{h+\zeta}\,|\mathbf{u}|\,\mathbf{u}, \tag{4.18}$$

where $\boldsymbol{f}$ is Coriolis frequency f multiplied by the unit vertical vector, h is mean water depth, ζ is deviation of free surface from mean depth, and C_d is bed-roughness coefficient associated with a quadratic drag law. If the rigid-lid approximation is made, i.e. $\zeta \ll h$, a very reasonable approximation given the local spatial scale of the turbine compared to the tidal wave length, the accompanying continuity equation reduces to $\nabla \cdot \mathbf{u} = 0$ and the Coriolis vector vanishes from the vorticity equation (as follows). By subsequently linearising the bottom friction, $(C_d/(h+\zeta))\,|\mathbf{u}|\,\mathbf{u} \rightarrow C_{\mathrm{L},d}\mathbf{u}$, and taking the curl of (4.18) the resulting vorticity equation obtained in GC13 gives

$$\frac{\partial \nabla^2\psi}{\partial t} + J(\psi, \nabla^2\psi) = -C_{\mathrm{L}}\nabla^2\psi - \nabla C_{\mathrm{L}} \cdot \nabla\psi, \tag{4.19}$$

where ψ is the streamfunction defined as $\mathbf{u} = (-\partial\psi/\partial y,\ \partial\psi/\partial x)$ and J is the Jacobian. At steady-state, neglecting the nonlinear material derivative and using polar coordinates (r, θ), the solution for the streamfunction is (GC13)

$$\psi = \begin{cases} -\left(1 - \frac{C_{\mathrm{L},T}}{C_{\mathrm{L},T} + 2C_{\mathrm{L},d}} \frac{R^2}{r^2}\right) u_0 r \sin\theta, & \text{for } r > R \quad (4.20) \\ -\frac{2C_{\mathrm{L},d}}{C_{\mathrm{L},T} + 2C_{\mathrm{L},d}} u_0 r \sin\theta, & \text{for } r \leq R. \quad (4.21) \end{cases}$$

where $C_{\mathrm{L},d}$ denotes the linear background friction ($C_\mathrm{L} = C_{\mathrm{L},d}$ for $r > R$) and $C_{\mathrm{L},T}$ is the additional friction associated with the turbine farm ($C_\mathrm{L} = C_{\mathrm{L},d} + C_{\mathrm{L},T}$ for $r \leq R$), as illustrated in Fig. 4.1b with streamlines shown as blue dashed lines. The streamfunction within the farm (4.21) is equivalent to uniform flow in the x-direction at constant speed, $u_T = 2u_0 C_{\mathrm{L},d}/(C_{\mathrm{L},T} + 2C_{\mathrm{L},d})$. Higher background friction $C_{\mathrm{L},d}$ invariably has the effect of directing a larger proportion of the flow through the farm such that the flow velocity increases within the farm ($r \leq R$) with $C_{\mathrm{L},d}$. Power dissipated by the turbines is given by the integral over the fluid of the linear friction coefficient of the turbines $C_{\mathrm{L},T}$ multiplied by the square of the flow speed within the farm (CG13)

$$P_{\mathrm{GC13}} = P_0 \lambda_T \left(\frac{\lambda_0}{\lambda_T + 2\lambda_0}\right)^2, \tag{4.22}$$

where $P_0 = 4\pi C_{0,\mathrm{ref}} \rho \pi R^2 u_0^3$, the non-dimensional background friction is $\lambda_0 = C_{\mathrm{L},d} h / C_{0,\mathrm{ref}} u_0$, and the non-dimensional turbine friction is $\lambda_T = C_{\mathrm{L},T} h / C_{0,\mathrm{ref}} u_0$. These are analogous, but not equivalent, to their counterparts for GC05 and V10. In order to facilitate comparison with the fully-spanned channel, λ_0 and λ_T are scaled by a non-stochastic reference drag coefficient $C_{0,\mathrm{ref}}$, so that typical values of λ_0 and λ_T are $O(1)$. The deterministic power extracted is maximised at a turbine drag of $\lambda_T^* = 2\lambda_0$.

As for the previous two models, uncertainty is introduced to background friction by expressing λ_0 as a normally distributed random variable with an expected value of μ_{λ_0} and variance $\sigma_{\lambda_0}^2$. Provided the variation is small compared to the mean, the power produced by the turbines may be expressed in terms of λ_0 by expanding (4.22) as a Taylor series in $\Delta\lambda_0 = \lambda_0 - \mu_{\lambda_0}$ about the deterministic case ($\Delta\lambda_0 = 0$).

Expected Power Performing the Taylor series expansion and evaluating the expectation operator for the leading-order effect of uncertainty, gives

$$\frac{1}{P_0}\mathrm{E}[P_{\mathrm{GC13}}] = \frac{\lambda_T \mu_{\lambda_0}^2}{(\lambda_T + 2\mu_{\lambda_0})^2} + \frac{{\lambda_T}^2(\lambda_T - 4\mu_{\lambda_0})}{(\lambda_T + 2\mu_{\lambda_0})^4}\sigma_{\lambda_0}^2 + \mathcal{O}\left(\mathrm{E}[\Delta{\lambda_0}^3]\right), \tag{4.23}$$

where the first term corresponds to the deterministic power (evaluated at mean background friction) and the second term provides a correction resulting from the background friction uncertainty. Figure 4.6 shows expected power as a function of turbine friction λ_T for different values of mean background friction coefficient μ_{λ_0} and standard deviation σ_{λ_0}. Unlike the quasi-steady limit of the fully-spanned channel (GC05), where the change in expected power from deterministic power is positive regardless of turbine drag, the sign of the correction term now depends on the rela-

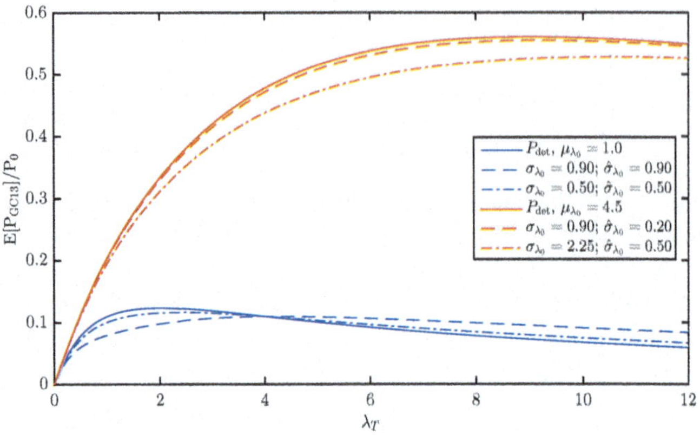

Fig. 4.6 Expected power produced by a laterally unconfined turbine farm (GC13) for two scenarios with mean background friction parameter values of $\mu_{\lambda_0} = 1.0$ and $\mu_{\lambda_0} = 4.5$. Deterministic power is indicated by solid lines. The dashed lines show the expected power from the two channels at the same value of standard deviation in the background friction parameter $\sigma_{\lambda_0} = 0.90$. The dot-dashed lines have the same value of relative standard deviation $\hat{\sigma}_{\lambda_0} = \sigma_{\lambda_0}/\mu_{\lambda_0} = 0.50$

tive magnitude of turbine drag and bed friction: for $\lambda_T < 4\lambda_0$ the expected power is reduced, and vice versa for $\lambda_T > 4\lambda_0$.

This non-monotonicity can be explained as follows. For sufficiently small values of background friction, power is approximately quadratic in λ_0 (because $P_{GC13} \propto {u_T}^2$ and $u_T \propto \lambda_0$ for $\lambda_0 \ll \lambda_T$) and a small increase $\lambda_0^+ > 0$ produces a greater increase in power than the reduction in power resulting from a decrease $\lambda_0^- < 0$ of the same magnitude (i.e. P_{GC13} is a convex function of λ_0). Hence a net increase in expected power occurs as a result of uncertainty, as may be seen in Fig. 4.4a for large $\lambda_T/\mu_{\lambda_0}$ (corresponding to small μ_{λ_0}). On the other hand, at large values of λ_0 a small increase in background friction has a relatively smaller effect on flow rate (cf. $u_T/u_0 \to 1$ for $\lambda_0 \gg \lambda_T$), and consequently power, than a decrease in background friction of equal magnitude. As λ_0 increases, the flow speed initially increases, but then tends towards a constant value. The decreasing rate of change of flow speed with λ_0 results in a concave dependence of power on λ_0 ($\partial^2 P_{\mathrm{GC13}}/\partial\lambda_0^2 < 0$) for sufficiently large λ_0. This results in a net decrease in expected power. The transition between the two regimes occurs at $\lambda_T = 4\lambda_0$, as is evident from (4.23).

In a completely-spanned channel (GC05), the flow rate decreases with increasing background friction (cf. $Q' \propto 1/\sqrt{\lambda_0 + \lambda_T}$), and the decreasing rate at which it does so (for increasing λ_0), corresponding to the flow being completely blocked, leads to convexity and a corresponding increase in expected power ($\partial^2 P_{\mathrm{GC05}}/\partial\lambda_0^2 > 0$). For a laterally unconfined turbine (GC13), the flow rate through the farm initially increases with increasing background friction (cf. $u_T/u_0 = 2\lambda_0/(\lambda_T + 2\lambda_0)$, but must do so at a decreasing rate (for increasing λ_0), because the flow through the farm cannot be stopped, leading to concavity and a corresponding decrease in expected power ($\partial^2 P_{\mathrm{GC05}}/\partial\lambda_0^2 < 0$). Examining Fig. 4.4b once more, as the number of turbines relative to the background friction (λ_T/λ_0) is increased, a transition occurs from concavity ($\partial^2 P_{\text{GC13}}/\partial\lambda_0^2 < 0$), associated with a reduction in expected power, to convexity ($\partial^2 P_{\text{GC13}}/\partial\lambda_0^2 > 0$), associated with an increase in expected power. From now on, the thesis refers to $\lambda_T < 4\lambda_0$ as background friction dominated and $\lambda_T > 4\lambda_0$ as turbine friction dominated.

Optimal Turbine Drag An analytical expression for the optimal turbine drag ${\lambda_T}^*_{\text{stoch}}$ can be found by maximising (4.23) with respect to λ_T such that

$$ {\lambda_T}^*_{\text{stoch}} = 2\mu_{\lambda 0}\left[1 + \frac{1}{2}\hat{\sigma}^2_{\lambda_0}\right] + \mathcal{O}(\text{E}[\Delta{\lambda_0}^3]) \quad \text{with} \quad \hat{\sigma}_{\lambda_0} = \frac{\sigma_{\lambda_0}}{\mu_{\lambda 0}}. \tag{4.24} $$

Compared to the downward shift in optimal turbine drag in response to background friction uncertainty for a fully-spanned channel (4.8) (GC05), which was only diminished in magnitude by the effect of inertia (V10), the optimal turbine drag in a laterally unconfined channel is shifted upwards (see Fig. 4.4b). This can be explained by referring to (4.9), noting that $P_{\lambda_T\lambda_T} < 0$ for optimum power. From Fig. 4.4a it is evident that at the deterministic optimum $\lambda_T/\lambda_0 = 2$ about which the perturbation is performed, increasing the number of turbines (λ_T) acts to reduce the concavity of the power with respect to background friction ($P_{\lambda_0\lambda_0\lambda_T} > 0$ at $\lambda_T/\lambda_0 = 2$) and thus $\Delta\lambda_T^* > 0$ from (4.9). It is optimal to move more into the turbine friction dominated regime.

Uncertainty in Power The variance in power to leading-order is given by

$$ \begin{aligned} \sigma_P^2 &= \frac{4\lambda_T^4\mu_{\lambda_0}^2}{(\lambda_T + 2\mu_{\lambda_0})^6}\sigma^2_{\lambda_0} + \mathcal{O}\left(\text{E}[\Delta{\lambda_0}^3]\right) \\ &= \left(P_{\text{GC13}}(\lambda_0 = \mu_{\lambda_0})\right)^2 \frac{4(\lambda_T/\mu_{\lambda_0})^2}{\left(2 + (\lambda_T/\mu_{\lambda_0})\right)^2}\hat{\sigma}^2_{\lambda_0} + \mathcal{O}\left(\text{E}[\Delta{\lambda_0}^3]\right), \end{aligned} \tag{4.25} $$

which is illustrated in Fig. 4.5. As can be seen from this figure and (4.25), increasing turbine drag as a share of mean background friction ($\lambda_T/\mu_{\lambda_0}$) increases the variability in power, tending towards a constant multiple of $\hat{\sigma}_{\lambda_0}$ as $\lambda_T/\mu_{\lambda_0} \to \infty$.

4.2.1 Comparison Between Models

Table 4.1 summarizes the effects of background friction uncertainty on expected power, optimal turbine drag, and power uncertainty for a fully-spanned channel in the quasi-steady limit (GC05) and with inertia (V10), and for a laterally unconfined turbine farm (GC13). As may be seen, bottom friction uncertainty acts to increase the expected power in a fully-spanned channel, but generally has an opposite effect in laterally unconfined farms. The optimal number of turbines with bottom friction uncertainty is lower in a fully-spanned channel and higher in laterally unconfined

Table 4.1 Leading-order effects of uncertainty in background friction ($\hat{\sigma}_{\lambda_0} = \sigma_{\lambda_0}/\mu_{\lambda_0}$) on the relative change in expected power, the relative standard deviation of power and the relative change in optimal turbine friction in drag-dominated (CG05) and inertia-dominated (V10) fully-spanned channels and in a laterally unconfined farm (GC13)

	Change in expected power	Standard deviation in power	Optimal turbine friction
	$(\mathrm{E}[P] - P_{\text{det}})\,/P_{\text{det}}$	σ_P/P_{det}	$\Delta\lambda_T^*/\lambda_{T,\text{det}}^*$
Drag-dominated fully- spanned channel (GC05)	$\frac{15}{8}\left(1+\frac{\lambda_T}{\mu_{\lambda_0}}\right)^{-2}\hat{\sigma}_{\lambda_0}^2$	$\frac{3}{2}\left(1+\frac{\lambda_T}{\mu_{\lambda_0}}\right)^{-1}\hat{\sigma}_{\lambda_0}$	$-\frac{5}{6}\hat{\sigma}_{\lambda_0}^2$
Inertia-dominated fully- spanned channel (V10)	Negative for $\mu_{\lambda_0} < 0.495$, positive for $\mu_{\lambda_0} > 0.495$	Smaller than GC05	Negative, but smaller in magnitude than GC05
Laterally unconfined farm (GC13)	$-\frac{\frac{\lambda_T}{\mu_{\lambda_0}}\left(4-\frac{\lambda_T}{\mu_{\lambda_0}}\right)}{\left(\frac{\lambda_T}{\mu_{\lambda_0}}+2\right)^2}\hat{\sigma}_{\lambda_0}^2$	$2\frac{\lambda_T}{\mu_{\lambda_0}}\left(2+\frac{\lambda_T}{\lambda_{C_d}}\right)^{-1}\hat{\sigma}_{\lambda_0}$	$+\frac{1}{2}\hat{\sigma}_{\lambda_0}^2$

farms. Bypass flow fundamentally changes how the system behaves under uncertainty. In fully-spanned channels, inertia acts to reduce the effect of uncertainty in background friction (V10 vs. GC05).

4.3 Calibration of Bottom Friction Uncertainty

In order to quantify its effect on power, it is necessary to estimate the magnitude of background friction uncertainty in the form of the relative standard deviation $\hat{\sigma}_{\lambda_0} = \sigma_{\lambda_0}/\mu_{\lambda_0}$. In each of the foregoing models, λ_0 is simply a linear function of the respective bottom roughness coefficients, assuming the effect of exit separation is negligible compared to that of bed roughness. This linearity means that the mean and standard deviation of C_d are scaled by the same constant to give the mean and standard deviation of λ_0. Consequently, the relative standard deviations may be set to be equal, i.e. $\sigma_{\lambda_0}/\mu_{\lambda_0} = \sigma_{C_d}/\mu_{C_d}$, where μ_{C_d} is the mean and σ_{C_d} the standard deviation of the bottom roughness coefficient. With a priori knowledge of both the tidal elevation and flow rate of a channel, the bed roughness coefficient could be relatively accurately determined from the phase difference between the two. With the exception of measurement errors, the uncertainty associated with background friction would be small, provided the flow conditions and thus the background friction experienced are not substantially altered by the introduction of turbines. In the absence of knowledge of both the tidal elevation and the flow rate, C_d is essentially unknown, as an observed elevation may be the result of an enormous number of combinations of bed roughness coefficients and flow rates. Equally, various values of tidal elevation and bed roughness may be combined to give an observed flow rate. Data on tidal elevation are often available. However, volumetric flux is usually far more difficult to determine. Point measurements of current velocities are sometimes available from Acoustic Doppler current profiler (ADCP) deployments and help confine the possible values of bed roughness coefficient to a region.

Two sources of background friction uncertainty are distinguished. First, the roughness length parameter Z_0 captures the magnitude of the friction coefficient at a site, which is dependent on the bed material and type and may be unknown, as well as varying across a given site. Z_0 is defined as height above the sea bead at which the flow speed theoretically becomes zero and is used as input in (semi-)empirical models to determine the bed roughness coefficient (see below). This thesis refers to the probability that the roughness length Z_0 is smaller than or equal to a value z_0, namely $\Pr(Z_0 \leq z_0)$, as arising from parametric uncertainty. Second, for a known value of the roughness length parameter z_0, many different models predict different friction coefficients C_d. The probability that the predicted friction coefficient C_d is smaller than or equal to a value c_d for a known value of the roughness length parameter z_0, namely $\Pr(C_d \leq c_d | Z_0 = z_0)$, arises from model uncertainty, which is conditional on parametric uncertainty. The unconditional probability $\Pr(C_d \leq c_d)$ then describes the likelihood that a particular value of c_d correctly captures the bed shear-stress due to the flow, and results from the two underlying sources of uncertainty, which are estimated separately below. In the following, capital variables refer to random variables, while lower-case variables to specific values that the random variables may take.

4.3.1 Parametric Uncertainty: $\Pr(\hat{Z}_0 \leq \hat{z}_0)$

The bed friction coefficient at a site is usually expressed as a function of relative roughness $\hat{Z}_0 = Z_0/h$, where Z_0 is the roughness length associated with a particular bed material and type and h is the water depth. Several phenomena contribute to the roughness length. These are the skin friction, due to the surface roughness of the sediment grains of the bed; the form drag, caused by the pressure field due to the presence of larger bed features; and the sediment-transport contribution, produced by momentum transfer of the flow to mobilised sediment particles [13]. An additional component relating to vegetation contributes in cases where there is plant growth at the bed. These components are commonly summed to give the total roughness length, i.e. $Z_{\text{tot}} = Z_{\text{skin}} + Z_{\text{form}} + Z_{\text{mom.trans.}} + Z_{\text{veg}}$ [13]. In absence of velocity profile measurements at a site, the bed roughness length may be estimated from knowledge of the bed conditions. The uncertainty in Z_0, then, stems from the difficulty in defining a single value for the roughness length due to spatial heterogeneity of the seafloor, variation in bed-grain sizes, change of bedforms with time (e.g. sand dunes travelling with the flow), as well as dependence of Z_0 on the hydrodynamic regime (i.e. whether the flow is hydrodynamically rough, smooth, or transitional).

Table 4.2 Parametric uncertainty resulting from variation in bed roughness length z_0 for different bed conditions. The number of values reported for different bed conditions and the resulting mean and variation factor are taken from [11]. The standard deviation of the natural logarithm of z_0 is given by the logarithm of the variation factor v.f. ($\sigma_{\ln z_0} = \ln(\text{v.f.})$). The (relative) standard deviation of bed roughness length $\hat{\sigma}_{z0}$ (σ_{z0}/μ_{z0}) were computed from these values and using the properties of the log-normal distribution

Bed material and type	No. of values reported	μ_{z0} (mm)	Variation factor, v.f.	σ_{z0} (mm)	σ_{z0}/μ_{z0}
Mud	1	0.2	–	–	–
Mud/sand	3	0.7	4.1	1.8	2.5
Silt/sand	1	0.05	–	–	–
Unrippled sand	7	0.4	2.0	0.3	0.8
Rippled sand	6	6.0	1.3	1.6	0.3
Sand/shell	2	0.3	4.5	0.9	2.9
Sand/gravel	7	0.3	6.7	1.8	6.0
Mud/sand/gravel	2	0.3	3.0	0.5	1.5
Gravel	4	3	1.6	1.5	0.5

In order to obtain an estimate for uncertainty in relative roughness $\hat{Z}_0$, the skin friction component of the bed roughness length is specifically considered. Table 4.2 lists values for the roughness lengths, obtained by fitting logarithmic velocity profiles for a range of different bed conditions, taken from Soulsby [11]. For seven of the nine bed conditions listed, [11] reports the geometric mean and variation factor obtained from a number of values reported in the literature. From these values, shown in Table 4.2, the standard deviation σ_{z_0} and the relative standard deviation σ_{z_0}/μ_{z_0} are computed. It is evident, that the uncertainty here is very considerable. Furthermore, the relative standard deviation depends strongly on bed type and ranges from 0.5 for gravel to 6.0 for a sand/gravel mixture. Because finer grains fill gaps between coarser grains, beds made up of a mixture of grain sizes have relatively low roughness lengths [11], while also exhibiting a higher standard deviation because the degree of filling will likely vary greatly according to the proportions of the different grain sizes present. This may be seen from the values in Table 4.2, where the relative standard deviation is typically larger for bed type mixtures than for beds made up of a single type.

When considering how to apply results such as those shown in Table 4.2 to a site, several scenarios in terms of available information and associated uncertainty are possible. Two limiting scenarios of these are considered. The first scenario is where accurate knowledge of the bed conditions exists, such that the relevant value of relative standard deviation σ_{z_0}/μ_{z_0} in the final column of Table 4.2 may be used. This presents a lower limit on uncertainty, identical to that of conditional model uncertainty in Sect. 4.4.1. In the second scenario, the bed conditions may be entirely unknown or might vary across a site. Assuming this latter limit, which forms a more realistic estimate, it is possible to proceed in a somewhat ad hoc fashion and

Table 4.3 Eight different formulae for calculating bed roughness coefficient C_d, used to estimate model uncertainty for given relative roughness $\hat{z}_0 = z_0/h$, taken from [12]. Here κ is von Kármán's constant and its values and those of the parameters α, β and B have been obtained from experimental and numerical data by different authors. The symbols used by Colebrook and White [2] are: k_s for the Nikuradse roughness, ν for viscosity, and u_* for friction velocity

Formula	Label in Fig. 4.7 and description	Parameters
$C_d = \left[\frac{\kappa}{B+\ln(\hat{z}_0)}\right]^2$	(1) - Deep water [12] Boundary-layer thickness δ	$\kappa = 0.40$
		$B = (\delta/2h) - \log(\delta/2h)$
	(2) - [2] $z_0 = (k_s/30) + (\nu/9u_*)$	$\kappa = 0.405$
		$B = 0.71$
	(3) - Full-depth logarithmic [12] velocity profile	$\kappa = 0.40$
		$B = 1$
$C_d = \alpha\hat{z}_0^{\beta}$	(4) - Manning-Strickler [12]	$\alpha = 0.0474$
		$\beta = 1/3$
	(5) - Dawson-Johns [3]	$\alpha = 0.0190$
		$\beta = 0.208$
	(6) - [11]	$\alpha = 0.0415$
		$\beta = 2/7$

assign equal probabilities to each of the bed conditions in Table 4.2 except for *mud* (only a single value reported), *silt/sand* (only a single value reported) and *rippled sand* (includes components of form drag and hence omitted) to estimate the relative standard deviation as follows

$$\hat{\sigma}_{z_0} = \frac{\sqrt{\sum_i w_i \sigma^2_{z_0,i}}}{\sum_i w_i \mu_{z_0,i}}, \tag{4.26}$$

where the subscript i corresponds to a row in Table 4.2, and weights are assigned according to the number of values reported (from Soulsby [11]) so that $\sum_i w_i = 1$. From this, a relatively large value of $\hat{\sigma}_{z_0} = 1.6$ is obtained, which is used as the base case. Ignoring uncertainty in the water depth, the relative standard deviation is set as $\hat{\sigma}_{\hat{z}_0} = \hat{\sigma}_{z_0} = 1.6$

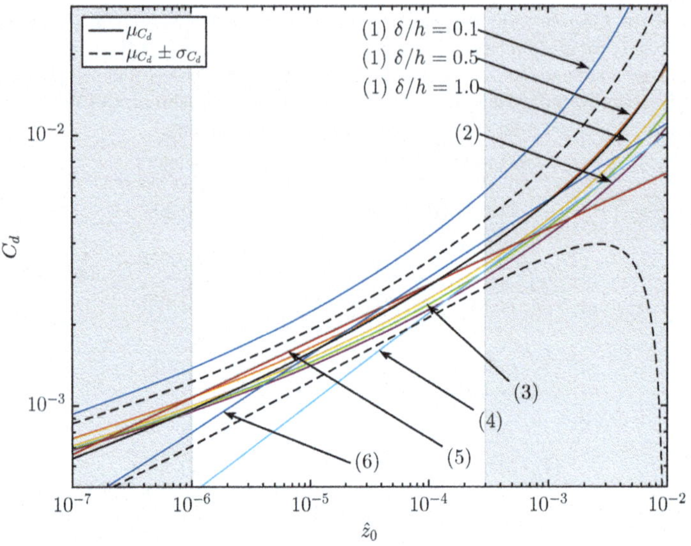

Fig. 4.7 Model uncertainty based on 8 different methods (numbered according to Table 4.3) to determine the drag coefficient C_d as a function of relative roughness $\hat{z}_0$, showing the the average $\mu_{C_d}(\hat{z}_0)$ (continuous black) and one standard deviation either side $\mu_{C_d}(\hat{z}_0) \pm \sigma_{C_d}(\hat{z}_0)$ (dashed black) as a function of $\hat{z}_0$. Shaded regions denote values of relative roughness beyond the limits for $\hat{z}_0$ relevant for tidal energy, i.e. $\hat{z}_0 = 1 \times 10^{-6}$ as lower and $\hat{z}_0 = 3 \times 10^{-4}$ as upper limit. Here h is the water depth and δ the boundary-layer thickness (see Table 4.3)

4.4 Bed Roughness Coefficient Models

Table 4.3 lists several commonly encountered formulae for bed roughness coefficients, derived from empirical and numerical experiments, as a function of relative roughness length $\hat{z}_0$. The formulae are shown in Fig. 4.7 to illustrate the spread in values of C_d for a given value of roughness length. The mean bed roughness coefficient μ_{C_d} and one standard deviation $\pm\sigma_{C_d}$ either side are superimposed onto the models.

4.4.1 Model Uncertainty: $\mathbf{Pr}(C_d \leq c_d | \hat{Z}_0)$

To estimate uncertainty resulting from the application of different friction coefficient models for a known value of the roughness length $\hat{z}_0$, eight different C_d-models summarised in Fig. 13 of Soulsby [12] (reproduced in Fig. 4.8a) are considered. The eight empirical models are derived from fitting experimental data to either a power-law relationship of the form $C_d = \alpha \hat{z}_0^{\beta}$ or a logarithmic law of the form $C_d = \left[\kappa / \left(B + \ln \hat{z}_0\right)\right]^2$, where κ is von Kármán's constant. Table 4.3 lists these two

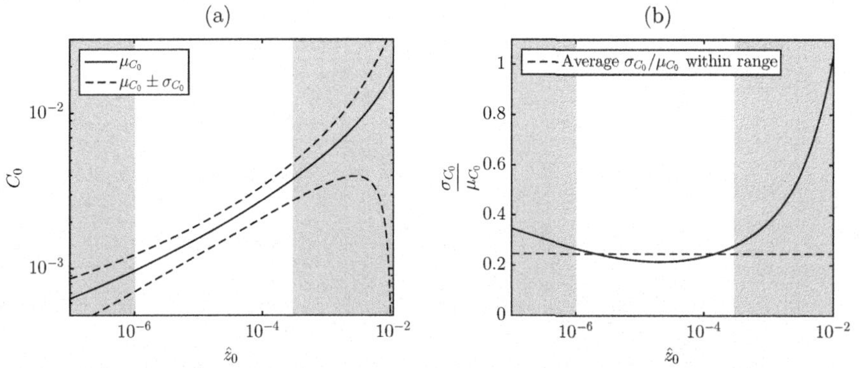

Fig. 4.8 **a** Model uncertainty based on 8 different methods to determine the drag coefficient C_d as a function of relative roughness $\hat{z}_0$, showing the average $\mu_{C_d}(\hat{z}_0)$ (continuous black) and one standard deviation either side $\mu_{C_d}(\hat{z}_0) \pm \sigma_{C_d}(\hat{z}_0)$ (dashed black) as a function of $\hat{z}_0$. **b** Relative standard deviation in C_d as a function of $\hat{z}_0$ (continuous) and the average (dashed line) for the range of consideration. Shaded areas denote values of relative roughness beyond the limits of $\hat{z}_0$ relevant for tidal energy, i.e. $\hat{z}_0 = 1 \times 10^{-6}$ as the lower limit and $\hat{z}_0 = 3 \times 10^{-4}$ as the upper limit

commonly used, empirical formulae for estimating C_d (left-hand column) and the values of the parameters α, β, B and κ fitted from experimental and numerical data by different authors. Taking an agnostic approach, equal weights are assigned to each of the eight models to determine the mean friction coefficient μ_{C_d} and the standard deviation σ_{C_d} across a range of values for $\hat{z}$ known with certainty. It is evident from Fig. 4.8a that model uncertainty is considerable.

It is useful to focus on the behaviour of the relative standard deviation across $\hat{z}_0$ values that are appropriate for tidal stream energy assessments. A lower bound on the $\hat{z}_0$ range in tidal channels is found by dividing the smallest roughness length, that for silt/sand ($z_0 = 5 \times 10^{-5}$ m), by a value of water depth typical for deep channels of approximately 50 m, giving a value of $\hat{z}_{\text{lower}} \approx 1 \times 10^{-6}$. An upper bound is found by dividing the largest value for z_0 (that for rippled sand, $z_0 = 6 \times 10^{-3}$ m) by a typical lower value for water depth of approximately 20 m, thus giving a value of $\hat{z}_{\text{upper}} \approx 3 \times 10^{-4}$. By considering the relative standard deviation throughout this (unshaded) range in Fig. 4.8b, it can be shown that this property has a weak dependence on the value of $\hat{z}_0$. The relative standard deviation $\hat{\sigma}_{C_d} \equiv \sigma_{C_d}/\mu_{C_d}$ varies between a minimum value of 0.21 and a maximum of 0.28, with an average value of 0.25 (indicated as a dashed line in Fig. 4.8b). At the midpoint of the range considered, $\hat{z}_0 = 1.51 \times 10^{-4}$, the value for the relative standard deviation is 0.22. A normal, non-skewed, distribution is appropriate for model uncertainty, which is evident from Fig. 4.9a, which shows the empirical cumulative distribution function. This distribution is estimated by creating a sample population from selecting four (arbitrary, yet equally spaced) relative roughness values of $\hat{z} = [10^{-6}, 10^{-5}, 10^{-4}, 10^{-3}]$, scaled by their local means and standard deviations, as shown in Fig. 4.9a.

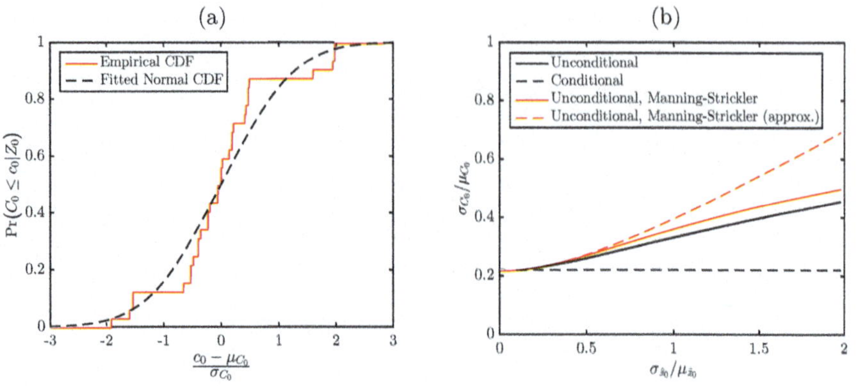

Fig. 4.9 Empirical cumulative distribution function for conditional model uncertainty (panel **a**) and variation of the relative standard deviation of the unconditional uncertainty in C_d with the relative standard deviation of parametric uncertainty at a mean relative roughness of $\mu_{\hat{z}_0} = 1.51 \times 10^{-4}$ (panel **b**)

4.4.2 Unconditional Uncertainty: $\Pr(C_d \le c_d)$

To address the scenario in which the bed conditions are not known or vary across a site, parametric uncertainty is combined with conditional model uncertainty from the previous sections, in order to obtain the unconditional uncertainty. Motivated by Soulsby [11], a log-normal probability distribution is used to capture parametric uncertainty and set its mean $\mu_{\hat{z}_0}$ equal to the average of the lower and upper bounds for $\hat{z}_0$ relevant to tidal energy determined earlier, namely $\mu_{\hat{z}_0} = 1.51 \times 10^{-4}$, for different values of $\hat{\sigma}_{z_0}$. The log-normal parametric uncertainty distribution is numerically convolved with the 8 equally weighted C_d-models from the previous section and the statistical moments calculated. Figure 4.9b shows the unconditional relative standard deviation $\hat{\sigma}_{C_d}$ as a function of relative standard deviation of relative roughness length $\hat{\sigma}_{\hat{z}_0}$ (continuous black line), the latter as a measure of parametric uncertainty. The conditional relative standard deviation $\hat{\sigma}_{C_d}$ is shown as a horizontal dashed black line and corresponds to the value of 0.22 obtained in the previous section.

In the case of a log-normal distribution for $\hat{z}_0$ and for the Manning-Strickler formula $C_d = \alpha\hat{z}_0^{\beta}$ (see Sect. 4.4), convolution may be achieved analytically. Assuming that α and $\hat{z}_0$ are independent random variables, the variance in C_d is given generally by

$$\mathrm{Var}[C_d] = \mathrm{Var}[\alpha]\mathrm{Var}[\hat{z}_0^{\beta}] + \mathrm{Var}[\alpha]\mathrm{E}[\hat{z}_0^{\beta}]^2 + \mathrm{E}[\alpha]^2\mathrm{Var}[\hat{z}_0^{\beta}], \tag{4.27}$$

which, by dividing by $\mathrm{E}[C_d]^2 = \mathrm{E}[\alpha]^2\mathrm{E}[\hat{z}_0^{\beta}]^2$, may be re-expressed as

$$\frac{\mathrm{Var}[C_d]}{\mathrm{E}[C_d]^2} = \frac{\mathrm{Var}[\alpha]}{\mathrm{E}[\alpha]^2}\frac{\mathrm{Var}[\hat{z}_0^\beta]}{\mathrm{E}[\hat{z}_0^\beta]^2} + \frac{\mathrm{Var}[\alpha]}{\mathrm{E}[\alpha]^2} + \frac{\mathrm{Var}[\hat{z}_0^\beta]}{\mathrm{E}[\hat{z}_0^\beta]^2}. \tag{4.28}$$

Or, in terms of relative variances,

$$\hat{\sigma}^2_{C_d,\mathrm{uncond}} = \hat{\sigma}^2_{C_d,\mathrm{cond}}\left[\hat{\sigma}^2_{\hat{z}_0^\beta} + 1\right] + \hat{\sigma}^2_{\hat{z}_0^\beta}, \tag{4.29}$$

where $\hat{\sigma}^2_{C_d,\mathrm{uncond}} \equiv \mathrm{Var}[C_d]/\mathrm{E}[C_d]^2$ denotes the unconditional variance of C_d, $\hat{\sigma}^2_{C_d,\mathrm{cond}} \equiv \mathrm{Var}[\alpha]/\mathrm{E}[\alpha]^2$ is the variance in C_d conditional on $\hat{z}_0$, and $\hat{\sigma}^2_{\hat{z}_0^\beta} \equiv \mathrm{Var}[\hat{z}_0^\beta]/\mathrm{E}[\hat{z}_0^\beta]^2$ is the relative variance of a power-law function of the uncertain bottom friction parameter. For a log-normally distributed random variable $\hat{z}_0$, the nth-order moment is given by $\mathrm{E}[\hat{z}_0^n] = \exp(n\mu_{\ln \hat{z}_0} + (1/2)n^2\sigma^2_{\ln \hat{z}_0})$, where $\mu_{\ln \hat{z}_0}$ and $\sigma^2_{\ln \hat{z}_0}$ are the mean and variance of the logarithm of $\hat{z}_0$, respectively. Hence the relative variance of $\hat{z}_0^\beta$ is given by $\hat{\sigma}^2_{\hat{z}_0^\beta} \equiv \mathrm{Var}[\hat{z}_0^\beta]/\mathrm{E}[\hat{z}_0^\beta]^2 = \mathrm{E}[\hat{z}_0^{2\beta}]/\mathrm{E}[\hat{z}_0] = \exp(\beta^2\sigma^2_{\ln \hat{z}_0})$. The relative variances of a log-normally distributed random variable and its logarithm are related by $\exp(\hat{\sigma}^2_{\ln \hat{z}_0})) = \hat{\sigma}^2_{\hat{z}_0} + 1$ and so $\hat{\sigma}^2_{\hat{z}_0^\beta} = (1 + \hat{\sigma}^2_{\hat{z}_0})^{\beta^2} - 1$. This expression is plotted in Fig. 4.9b as a continuous red line. It shows good agreement with the unconditional variance from numerically exact convolution (continuous black line); apparent disagreements are due to models of alternative form also being included in the curve (cf. Sect. 4.4). For small values of uncertainty in $\hat{z}_0$ (parametric uncertainty) and α (model uncertainty), (4.29) may be expanded in terms of $\hat{\sigma}^2_{\hat{z}_0}$ to give

$$\hat{\sigma}^2_{C_d,\mathrm{uncond}} \approx \hat{\sigma}^2_{C_d,\mathrm{cond}} + \beta^2\hat{\sigma}^2_{\hat{z}_0}, \tag{4.30}$$

where only leading-order terms are considered in both relative variances and their products are ignored. The dashed red line in Fig. 4.9b shows that (4.30) accurately represents (4.29), except for large values of $\hat{\sigma}_{\hat{z}_0}$.

At the base case value for parametric uncertainty of $\hat{\sigma}_{\hat{z}_0} = 1.6$ (derived from (4.26) and Table 4.2), an estimate is obtained for the unconditional relative standard deviation of $\hat{\sigma}_{C_d} = 0.41$ from Fig. 4.9b. This value $\hat{\sigma}_{\lambda_0} = \hat{\sigma}_{C_d} = 0.41$ is used in the next section to estimate the quantitative impact of bed roughness uncertainty. It should be emphasised that the estimates are indicative, not definite.

4.5 Quantitative Estimates of the Effect of Uncertainty

Expected Power Figure 4.10a shows the change in expected power as a percentage of deterministic power for the base case value of relative background friction uncertainty $\hat{\sigma}_{\lambda_0} = 0.41$. For a fully-spanned channel dominated by channel drag (GC05), such as a shallow, long channel with $\mu_{\lambda_0} \gg 1$, the increase in expected power can be as large as 30%. In fact, in the limit of very few turbines (λ_T), it is the case

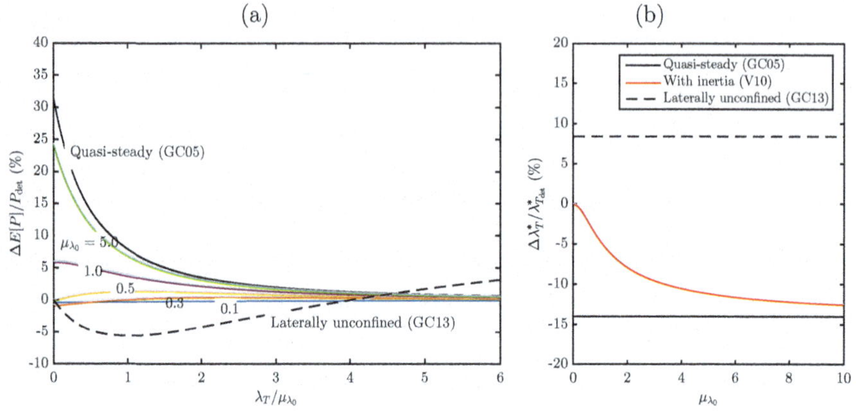

Fig. 4.10 Quantitative estimates of relative change in expected power due to background friction ($\hat{\sigma}_{\lambda_0} = 0.41$) as a function of turbine drag scaled with mean background friction at different values of μ_{λ_0} (**a**) and relative change in optimal turbine friction in the presence of uncertainty in the background friction parameter (**b**) for a fully-spanned tidal channel (GC05 and V10)

that $(\text{E}[P] - P_{\text{det}})/P_{\text{det}} = (15/8)\hat{\sigma}^2_{\lambda_0} \approx 32\%$ (cf. Table 4.1). However, in a deeper channel representative of the Pentland Firth ($\mu_{\lambda_0} = 1.0$, see Vennell et al. [17]) the increase in expected power would only be of the order of a few percent (6%) and would tend to reduce as more turbines are added. For laterally unconfined channels, the effects are negative and generally small (less than 5–10%).

Optimal Turbine Drag Figure 4.10b shows the change in optimal turbine drag as a percentage of the deterministic optimum for $\hat{\sigma}_{\lambda_0} = 0.41$. The change is between -14% for a drag-dominated fully-spanned channel (GC05) and $+8\%$ for a laterally

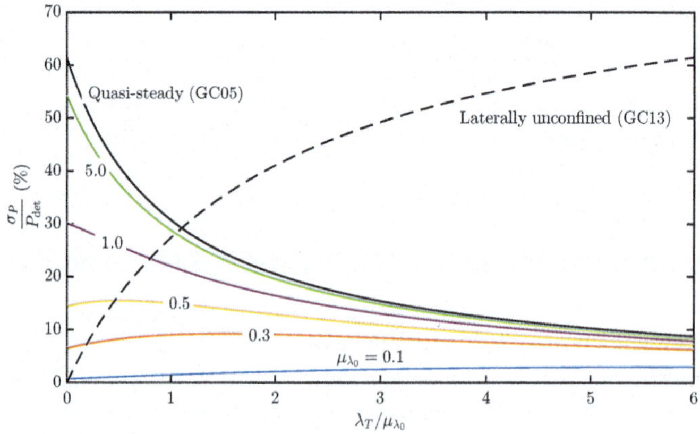

Fig. 4.11 Quantitative estimates of relative standard deviation in power due to background friction ($\hat{\sigma}_{\lambda_0} = 0.41$) as a function of turbine drag scaled mean with mean background friction and at different values of μ_{λ_0} for the laterally unconfined farm (V10)

unconfined channel (GC13). Inertia acts to reduce the decrease in optimal turbine drag for fully-spanned channels; a channel representative of the Pentland Firth ($\mu_{\lambda_0} = 1.0$), exhibits a reduction in the optimal turbine drag of 5%.

Uncertainty in Power Figure 4.11 shows the standard derivation of power as a percentage of the deterministic power for $\hat{\sigma}_{\lambda_0} = 0.41$. For drag-dominated fully-spanned channel (GC05) with few turbines, the relative standard derivation reaches 62%. Again, inertia reduces this. In a channel representative of the Pentland Firth ($\mu_{\lambda_0} = 1.0$), the relative standard deviation is estimated to be 30%.

4.6 Conclusions

Estimates of the tidal power that can be extracted at a given site are subject to significant uncertainty, with different estimates sometimes more than an order of magnitude apart. Of the many sources of uncertainty, uncertainty in bed friction can be considerable - both due to unknown and spatially-varying bed conditions and variation in the predictions of different bed friction models. To illustrate this, the parametric uncertainty is estimated resulting from lack of knowledge of bed conditions at a particular site, to be associated with a relative standard deviation of $\hat{\sigma}_{\hat{z}_0} = 1.6$, by assigning equal probabilities to a range of commonly occurring bed types. Even with precise knowledge of the bed conditions, a relative standard deviation is estimated of 0.22 associated with the range of predictions for C_d from different models outlined in Table 4.3. These uncertainties are combined to give an unconditional uncertainty in bed roughness of $\hat{\sigma}_{C_d} = 0.41$ (one relative standard deviation) related to a typical site for tidal turbine deployment. This estimate constitutes a lower limit for uncertainty in the bed roughness coefficient at a particular site for the data presented in Table 4.2, because it is assumed that the bed conditions (and their variability) are known. In reality, this knowledge is unlikely, and so the uncertainty in C_d is likely to be greater. Furthermore, spatio-temporal variability in bed conditions, which is not discussed here, will act to increase the value for $\hat{\sigma}_{C_d}$. For a given site, the uncertainty may be constrained by performing appropriate seabed surveys and ADCP measurements such that a better estimate for bed roughness coefficient may be found through additional data points being present in the model calibration. The certainty of this calibration, however, is limited because the spatial heterogeneity of the sea bed means that the application of a single bed roughness coefficient will not be able to give complete agreement with velocity vectors and sea level heights at all measurement locations.

In order to make a quantitative assessment of the effect of background friction uncertainty on estimates of tidal power potential, such uncertainty has been incorporated in three idealised models of tidal energy extraction, each of which captures a different element of the key physics. In Garrett and Cummins [5] an analytic solution is derived for the power potential of a channel in the drag-dominated limit and fully-spanned by turbines. Vennell [16] relaxes this limit by retaining inertia in the governing equation for a fully-spanned channel and derives an analytic solution for power to an approximate form of the governing equation. Finally, [6] allow for bypass flows by considering a laterally unconfined turbine farm. In particular, perturbation methods have been used to derive leading-order estimates for the effect of uncertainty in the value of bed roughness coefficient on three key quantities for each of the models: expected power, standard deviation in power and optimal turbine drag. In the presence of background friction uncertainty and nonlinearity in the model, evaluating power for the expected value of background friction does not give the same answer as evaluating the expectation of power for the distribution of values of background friction (cf. Jensen's inequality). It is the difference between the two that is considered here when comparing expected power with deterministic power (evaluated at the mean value of background friction). A similar issue is encountered in wind energy assessment, where the median is used as a measure of power under uncertainty because it is invariant under monotonic nonlinear transformations. Evidently, power is now a random variable and the standard deviation can be taken as a relevant measure of its distribution to understand confidence in tidal resource estimates. Finally, the turbine drag chosen to optimize expected power is different from that chosen to optimize deterministic power.

Our conclusions are as follows. First, for fully-spanned channels (GC05 and V10), two regimes can be identified. In the drag-dominated regime, the expected power is larger than the deterministic power, whereas in the inertia-dominated regime the opposite is true. Inertia has the effect of bounding the flow rate at low values of bed roughness, such that the increase in expected power is smaller and even reversed at sufficiently low total channel drag (background + turbine). For channels in which the flow may be diverted around the turbines (GC13), the expected power always decreases, except for extremely large turbine drag (i.e. very many turbines installed). Quantitatively, the analytic model shows that expected power can increase by as much as 32% for drag-dominated, quasi-steady channels, which are typically shallow and long, while reducing expected power by only 6% in laterally unconfined flow. In a channel representative of the Pentland Firth ($\mu_{\lambda_0} = 1.0$), the increase in expected power may only be of the order of a few percent.

Second, uncertainty in power is only enhanced compared to background uncertainty for very drag-dominated fully-spanned channels (and for laterally-unconfined channels with very large turbine drag). Inertia has the effect of reducing power uncertainty, because power becomes less sensitive to bottom drag in the presence of inertia, and variation in bed roughness produces a relatively smaller variation in power. Laterally-unconfined channels behave in the opposite way: for low values of turbine drag, less of the flow is diverted around the farm, and the flow rate tends towards a constant, independent of bed roughness. For a channel representative of the Pentland Firth, the uncertainty in removable power may be as large as 30% (one relative standard deviation) for small-scale turbine deployments. This value increases to over 50% for a small, high flow-rate channel ($\mu_{\lambda_0} = 4.5$), indicating that while the shift in expected power resulting from considering uncertainty may be negligible, variation in this power can be considerable. For example, the 95% confidence interval for the power from the Pentland Firth due to uncertainty in bed roughness will be at $\pm 2\hat{\sigma}_P = \pm 60\%$ of the mean power value determined. For a mean power of 5 GW (the mean of the range 0.62–9 GW given in the Introduction), then the range of likely values for power estimates from the Pentland Firth is 2–8 GW, spanning a significant portion of the range of reported values. However, it must be emphasised that the reported estimates are taken from different models, with different physical assumptions, containing sources of uncertainty other than bed friction (the focus of the present thesis) and which may contribute to a greater extent to the range of mean power estimates reported above.

Third, the turbine drag that maximises expected power in the presence of background uncertainty is greater compared to its deterministic value for laterally-unconfined channels (GC13) and smaller for fully-spanned channels (GC05), with uncertainty reducing the size of this effect (V10). Generally, however, this effect is small (between -8 and $+14\%$).

There are a number of limitations to these findings. First, the models considered herein are idealised and do not take into account the complex bathymetry of actual tidal sites and associated flow curvature, the complexity of the tidal forcing components, or deformation of the free surface. This limits the extent to which the findings from the models may be applied to real sites that exhibit such features. Furthermore, the models used are depth-averaged and so provide suitable power estimates only for regional scale energy extraction by large turbine deployments [1]. To take these into account and validate the predictions made in this chapter, Chap. 6 considers the effect of bed roughness uncertainty in a numerical model applied to the Pentland Firth. Second, this chapter has considered only the effect on removable power, which does not take into account mixing in the wake of the turbines, instead of available power. Future work, would apply the present methodology to linear momentum actuator disk theory [7] to include the effect of wake mixing. Third, the chapter has solely focused on uncertainty in background friction and not turbine drag itself. Future work would consider uncertainty regarding the correct value for enhanced bed roughness to use in a depth-averaged model to capture accurately the thrust exerted by rows of turbines [9, 18].

References

1. Adcock TAA, Draper S, Nishino T (2015) Tidal power generation - A review of hydrodynamic modelling. Proc Inst Mech Eng A 1–17
2. Colebrook CF, White CM (1937) Experiments with fluid friction in roughened pipes. Proc Royal Soc A 161:367–381
3. Dawson G, Johns B, Soulsby R (1983) A numerical model of shallow-water flow over topography. In: Johns B (ed) Physical oceanography of coastal and shelf seas, chapter 6. Elsevier, Amsterdam, pp 267–320
4. Draper S (2011) Tidal stream energy extraction in coastal basins. PhD thesis, University of Oxford
5. Garrett C, Cummins PF (2005) The power potential of tidal currents in channels. Proc Royal Soc A 461:2563–2572

6. Garrett C, Cummins PF (2013) Maximum power from a turbine farm in shallow water. J Fluid Mech 714:634–643
7. Houlsby GT, Draper S, Oldfield M (2008) Application of linear momentum actuator disc theory to open channel flow. Technical report, University of Oxford
8. Karsten RH, McMillan JM, Lickley MJ, Haynes RD (2008) Assessment of tidal current energy in the Minas Passage, Bay of Fundy. Proc Inst Mech Eng A 222:493–507
9. Kramer SC, Piggott MD (2016) A correction to the enhanced bottom drag parameterisation of tidal turbines. Renew Energy 92:385–396
10. Kreitmair MJ, Draper S, Borthwick AGL, van den Bremer TS (2019) The effect of uncertain bottom friction on estimates of tidal current power. R Soc open sci, 6
11. Soulsby RL (1983) The bottom boundary layer of shelf seas. In: Johns B (ed) Physical oceanography of coastal and shelf seas, vol 35. Elsevier, Amsterdam, pp 189–266
12. Soulsby RL (1990) Tidal-current boundary layers. In: Le Méhauté B, Hanes DM (eds) Ocean engineering science: the sea, vol 9, B, chapter 15. Wiley, New York, pp 523–567
13. Soulsby RL (1997) Dynamics of marine sands. Thomas Telford Publications, London, United Kingdom
14. Sutherland G, Foreman M, Garrett C (2007) Tidal current energy assessment for Johnstone Strait, Vancouver Island. Proc Inst Mech Eng A 221(2):147–157
15. Terra GM, van de Berg WJ, Maas LRM (2005) Experimental verification of Lorentz' linearization procedure for quadratic friction. Fluid Dyn Res 36(3):175–188
16. Vennell R (2010) Tuning turbines in a tidal channel. J Fluid Mech 663:253–267
17. Vennell R, Funke SW, Draper S, Stevens C, Divett T (2015) Designing large arrays of tidal turbines: a synthesis and review. Renew Sustain Energy Rev 41:454–472
18. Vogel CR, Willden RHJ, Houlsby GT (2013) A correction for depth-averaged simulations of tidal turbine arrays. In: Proceedings of the 10th European wave and tidal energy conference, Aalborg
19. Zimmerman JTF (1982) On the Lorentz linearization of a quadratically damped forced oscillator. Phys Lett A 89(3):123–124

Chapter 5
Uncertainty in Turbine Resistance

In order to incorporate the presence of tidal turbines, numerical models must implement the proper thrust produced by the turbines to ensure that correct amounts of momentum and energy are removed [5]. Several models (e.g. TELEMAC, MIKE 21, ROMS and FVCOM) incorporate the thrust due to arrays of turbines at regional scale, informed by properties determined from turbine-scale models, in terms of an equivalent drag term. For depth-averaged models, this equates to enhanced bed roughness [5]. In practice, the force exerted by turbines is often assumed to have quadratic dependence on the depth-averaged velocity. Within depth-averaged numerical models, the presence of turbines may be represented by an increase in the local bed friction coefficient at either a grid cell of similar size as the physical turbine or else smeared out over a larger area. There is, however, uncertainty in the correct value to use for this frictional increase. Arrangement of the tidal devices, the presence of support structures, coastal geometry surrounding the turbines [2], and modelling of the upstream velocity (dependent on numerical grid resolution) [5] all affect the value used for the drag coefficient representing the turbine thrust.

This chapter examines the effect of uncertainty in turbine drag representation on the average extractable power from a channel and the standard deviation of its distribution. The chapter also explores how the introduction of uncertainty alters the value of optimal turbine tuning used to maximise the expected power removed. To this end, the three analytic models of tidal power extraction are again considered (for further details see Chap. 4). A very brief recap follows. The first model is that of Garrett and Cummins [3] (hereafter GC05) wherein an analytic solution is derived for the power from a channel connecting two large ocean basins in the quasi-steady limit. The second is that of Vennell [6] (hereafter V10) who relaxes the quasi-steady assumption necessary for the analytic solution of GC05 and determines an approximate solution for power from a channel by linearising the drag term in the governing equation. In both cases, the channel is fully spanned by turbines. The third model

M. J. Kreitmair, *The Effect of Uncertainty on Tidal Stream Energy Resource Estimates*,
Springer Theses, https://doi.org/10.1007/978-3-030-57658-5_5

is from Garrett and Cummins [4] (hereafter GC13) in which an expression for the power from a circular turbine farm in unconfined flow is obtained. These models, despite idealisations, are nonetheless useful to understanding leading-order trends in power extraction from a channel [1]. In the models, solutions are derived in terms of a bed roughness parameter and a turbine drag parameter. Following the approach taken in Chap. 4, uncertainty in the latter is now introduced by performing a local expansion of the solution for power in terms of a perturbation to the turbine drag around its mean value. The resulting expression for expected power includes stochastic correction terms in addition to the deterministic term, where the correction terms are proportional to the statistical moments of the probability density function of the turbine drag.

5.1 Stochastic Expansion

Uncertainty is introduced into the GC05, V10, and GC13 models by expressing the turbine drag parameter as a random variable with a known mean value μ_{λ_T} and a random fluctuation $\Delta\lambda_T$ about this mean, i.e. $\lambda_T = \mu_{\lambda_T} + \Delta\lambda_T$. For fluctuations that are small compared to μ_{λ_T}, the above solutions for power may be expanded as a Taylor series in $\Delta\lambda_T$ about the mean turbine drag. For the GC05 model, expansion of (4.5) gives

$$\frac{1}{P_0}P_{\text{GC05}} = \frac{\mu_{\lambda_T}}{(\lambda_0 + \mu_{\lambda_T})^{3/2}} - \frac{\mu_{\lambda_T} - 2\lambda_0}{2(\lambda_0 + \mu_{\lambda_T})^{5/2}}\Delta\lambda_T + \frac{3}{8}\frac{\mu_{\lambda_T} - 4\lambda_0}{(\lambda_0 + \mu_{\lambda_T})^{7/2}}\Delta\lambda_T{}^2 + \mathcal{O}(\Delta\lambda_T{}^3). \tag{5.1}$$

From this expansion, and the equivalent for the models V10 and GC13, analytic expressions for expected power, standard deviation, and optimum mean turbine drag are determined in terms of the statistical moments of the turbine drag input distribution.

Expected Power Application of the expectation operator to (5.1) gives the expected power for the GC05 model:

$$\frac{1}{P_0}\mathrm{E}[P_{\text{GC05}}] = \frac{\mu_{\lambda_T}}{(\mu_{\lambda_T} + \lambda_0)^{3/2}} + \frac{3}{8}\frac{(\mu_{\lambda_T} - 4\lambda_0)}{(\mu_{\lambda_T} + \lambda_0)^{7/2}}\sigma_{\lambda_T}{}^2 + \mathcal{O}(\mathrm{E}[\Delta\lambda_T{}^3]), \tag{5.2}$$

where $\sigma_{\lambda_T}{}^2 = \mathrm{E}[\Delta\lambda_T{}^2]$ is the variance in turbine drag parameter. The first term in the expression is simply the power calculated at a turbine drag equal to the mean, $\lambda_T = \mu_{\lambda_T}$. The second term is a stochastic correction, taking into account the spread of the turbine drag parameter values about the mean with a standard deviation of σ_{λ_T}. Higher-order terms have been neglected here but give further contributions that

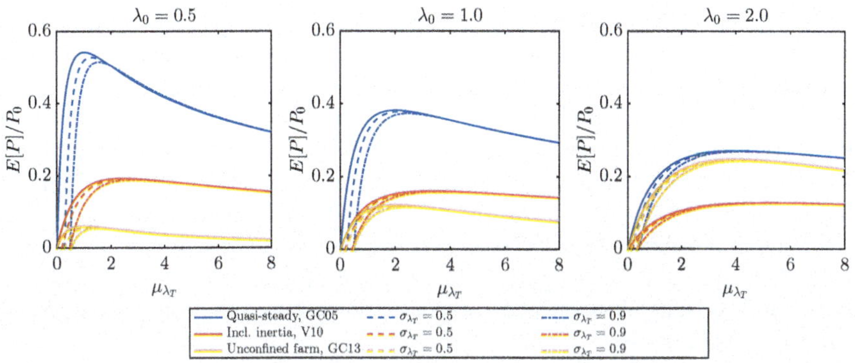

Fig. 5.1 Expected power as a function of mean turbine drag for different channels, $\lambda_0 = \{0.5, 1.0, 2.0\}$, and different values of turbine drag standard deviation. Power from deterministic models is shown as a continuous line. Dashed and dot-dashed lines represent expected power from stochastic expansions at $\sigma_{\lambda_T} = 0.5$ and $\sigma_{\lambda_T} = 0.9$ respectively

take into account e.g. asymmetry in the input distribution of the turbine drag (i.e. the skewness) or its peakedness (kurtosis).

Equivalently, the expected power from the GC13 model is given by

$$\frac{1}{P_0}\mathrm{E}[P_{\mathrm{GC13}}] = \frac{\lambda_0\mu_{\lambda_T}}{(\mu_{\lambda_T}+2\lambda_0)^2} + \frac{\lambda_0^2(\mu_{\lambda_T}-4\lambda_0)}{(\mu_{\lambda_T}+2\lambda_0)^4}{\sigma_{\lambda_T}}^2 + \mathcal{O}(\mathrm{E}[\Delta{\lambda_T}^3]), \tag{5.3}$$

and the same procedure may be used to find the expected power from V10. The expression for $\mathrm{E}[P_{V10}]$ has not been given here due to its length.

The expected power under turbine drag uncertainty for the models GC05, V10, and GC13 is plotted in Fig. 5.1 for different channels, i.e. different values of bed roughness parameter λ_0, and at different values of turbine drag volatility $\sigma_{\lambda_T} = 0.5$ and $\sigma_{\lambda_T} = 0.9$. The value $\lambda_0 = 0.5$ corresponds to a very large and deep tidal strait where inertia makes a significant contribution to the channel dynamics. A value of bed roughness parameter of $\lambda_0 = 1.0$ is representative of a channel similar to that of the Pentland Firth [7]. The value $\lambda_0 = 2.0$ indicates a small, shallow channel where natural bed roughness becomes dominant in the dynamic balance.

Irrespective of channel or model, for small values of μ_{λ_T}, the effect of uncertainty is to reduce the expected power relative to the deterministic solutions. Furthermore, the greater the standard deviation in turbine drag, the greater the reduction in expected power. However, as the value of μ_{λ_T} increases, a changeover occurs and the expected power becomes enhanced compared to the deterministic power. For the models of GC05 and GC13 this changeover occurs at $\mu_{\lambda_T} = 4\lambda_0$. The magnitude of the increase is small at such high values of turbine drag because power is increasingly insensitive to fluctuations in turbine drag at large total channel drag. However, it may be noted that such a high value of bed roughness parameter corresponds to an feasibly large deployment of turbines and is double the optimal turbine drag.

More generally, the stochastic correction terms in (5.2) and (5.3) arise from the third term in the respective Taylor expansions, which in turn is given by the second derivative of the solution for power. The leading-order change, therefore, is dependent on the concavity of the power curve with respect to the turbine drag. Near the optimum the second derivative of power with respect to turbine drag is necessarily negative. Hence, the stochastic correction term also is negative for sufficiently low values of turbine drag, $\mu_{\lambda_T} < 4\lambda_0$. That is, for low turbine deployments, realistic for early stages of turbine deployment, the effect of uncertainty in turbine drag will invariably be to decrease the expected power. As the value of μ_{λ_T} increases with respect to the natural bed roughness, change in expected power becomes positive as the power curve becomes convex.

The behaviour of the power can be observed more generally in response to changes in turbine deployment relative to the natural bed roughness. Figure 5.2 shows the fractional change in expected power per unit relative variance in λ_T as a function of the scaled mean turbine drag. For the models GC05 and GC13, the curves collapse onto a single curve upon scaling. For the V10 model, the behaviour of the expected power depends on the dynamic balance between natural friction and inertia in the channel and thus the value of λ_0, as shown by the set of coloured curves. Channels which are more inertia dominated, i.e. lower value of λ_0, remain concave for greater values of $\mu_{\lambda_T}/\lambda_0$ than the quasi-steady model of GC05 and hence show a decrease in expected power for a larger range of turbine deployment. Furthermore, expected power is reduced by a greater proportion in the V10 model than in the quasi-steady case as, while the absolute change in power is very similar for all three models, the deterministic power from V10 is significantly smaller than that of GC05.

Interesting to note is that the behaviour of the GC13 model under uncertainty is similar as that of GC05 and V10, despite describing very different turbine layouts.

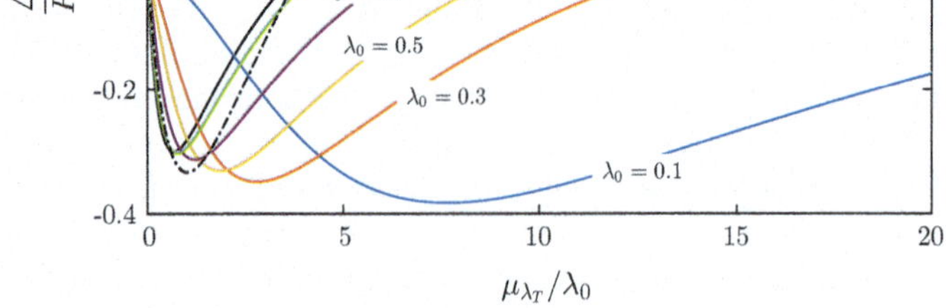

Fig. 5.2 Fractional change in expected power per unit relative variance in turbine drag as a function of mean turbine drag scaled by background friction. Coloured lines correspond to the model V10 at different values of bed roughness parameter λ_0

This is because the dependence of the power on the turbine drag is similar in all three models: there is a trade-off between the slowing of the flow with increasing thrust exerted by the turbines and the increased power extracted via the same thrust. As a result, any uncertainty in the turbine drag will have similar qualitative effects in all three models. Further, because the changeover of the power curve from concave to convex occurs at the same relative turbine drag $\mu_{\lambda_T}/\lambda_0$ for both GC13 and GC05, the switch from a negative to a positive change in expected power coincides for the two models at $\mu_{\lambda_T}/\lambda_0 = 4$. This similarity is in contrast to the effect on power due to uncertain bed roughness, for which the channel models, GC05 and V10, give very different results to the unconfined model of GC13 in Chap. 4. In GC13, the natural bed roughness acts to funnel the flow through the farm in addition to slowing it such that the effect of bed roughness uncertainty has the opposite effect in GC13 compared to GC05 and V10.

Standard Deviation The spread of possible power values about the mean power is given by the standard deviation σ_P and may be calculated from ${\sigma_P}^2 = \mathrm{E}\left[(P - \mathrm{E}[P])^2\right]$. To leading-order, for the GC05 model, the standard deviation in power is thus given by

$$\sigma_{P_{\mathrm{GC05}}} = \frac{1}{2}\frac{\left|\mu_{\lambda_T} - 2\lambda_0\right|}{(\mu_{\lambda_T} + \lambda_0)^{5/2}}\sigma_{\lambda_T} + \mathcal{O}(\mathrm{E}[\Delta{\lambda_T}^3]) \tag{5.4}$$

$$= \frac{1}{2}P_{\mathrm{GC05}}\frac{\left|\mu_{\lambda_T} - 2\lambda_0\right|}{\mu_{\lambda_T}(\mu_{\lambda_T} + \lambda_0)}\sigma_{\lambda_T} + \mathcal{O}(\mathrm{E}[\Delta{\lambda_T}^3]). \tag{5.5}$$

Similarly, for GC13, the standard deviation is then

$$\sigma_{P_{\mathrm{GC13}}} = \frac{\lambda_0^2\left|\mu_{\lambda_T} - 2\lambda_0\right|}{(\mu_{\lambda_T} + 2\lambda_0)^3}\sigma_{\lambda_T} + \mathcal{O}(\mathrm{E}[\Delta{\lambda_T}^3]) \tag{5.6}$$

$$= P_{\mathrm{GC13}}\frac{\left|\mu_{\lambda_T} - 2\lambda_0\right|}{\mu_{\lambda_T}(\mu_{\lambda_T} + 2\lambda_0)}\sigma_{\lambda_T} + \mathcal{O}(\mathrm{E}[\Delta{\lambda_T}^3]). \tag{5.7}$$

The ratio of standard deviation in power to standard deviation in turbine drag as a function of mean turbine drag is plotted in Fig. 5.3 for the same set of channels as Fig. 5.1. This ratio $\sigma_P/\sigma_{\lambda_T}$ is the amplification of the uncertainty in λ_T to uncertainty in power. For almost all values of mean turbine drag the amplification is less than unity, i.e. the models act to decrease input uncertainty. In the limit of very little turbine deployment $\mu_{\lambda_T} \to 0$, the amplification factors for GC05 and GC13 tend to $\sigma_{P_{\mathrm{GC05}}}/\sigma_{\lambda_T} \to \lambda_0^{-3/2}$ and $\sigma_{P_{\mathrm{GC13}}}/\sigma_{\lambda_T} \to 1/4$ respectively. For the model of V10, the amplification in this limit depends on the value of λ_0 but tends to $4/(3\pi)$ as the inertia increases $\lambda_0 \to 0$ and falls off with increasing bed roughness.

A general trend may be explored through considering the amplification of standard deviation, that is, the ratio of the relative standard deviation in power $\hat{\sigma}_P = \sigma_P/P_{\mathrm{det}}$ to the relative standard deviation in turbine drag $\hat{\sigma}_{\lambda_T} = \sigma_{\lambda_T}/\mu_{\lambda_T}$. This has been plotted as a function of the mean turbine drag relative to bed roughness parameter in Fig. 5.4.

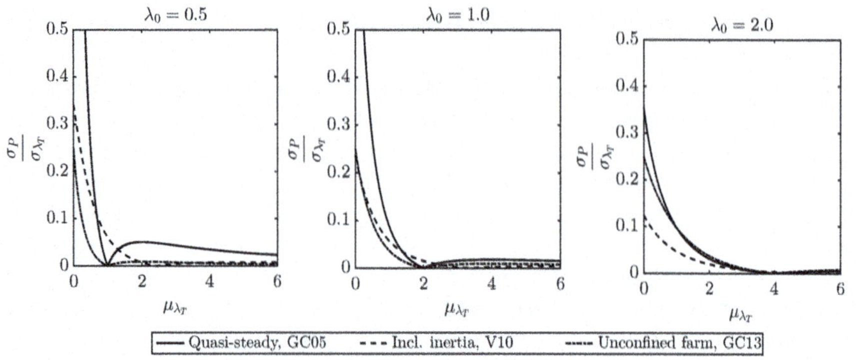

Fig. 5.3 Amplification of uncertainty in turbine drag to uncertainty in power measured by the ratio of standard deviation in power to standard deviation in turbine drag as a function of mean turbine drag for different channels, $\lambda_0 = \{0.5, 1.0, 2.0\}$

Fig. 5.4 Ratio of relative standard deviation in power to relative standard deviation in turbine drag, as a function of mean turbine drag scaled by bed roughness parameter

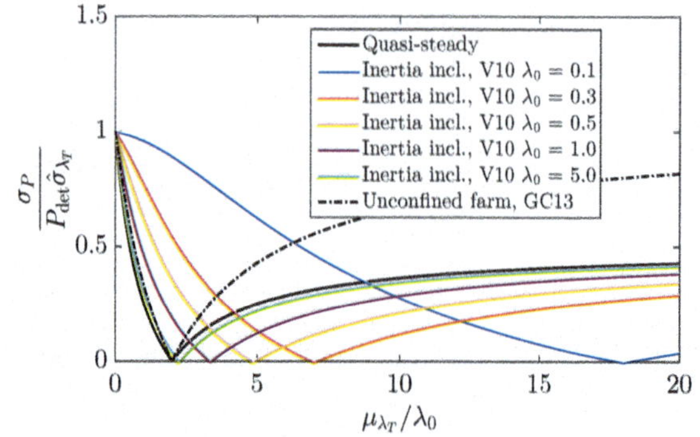

For all models, the ratio drops to zero for a particular value of μ_{λ_T}, corresponding to the optimum turbine drag of the deterministic solution. The standard deviation in power is related to the standard deviation in the uncertain turbine drag through the gradient in the power curve as the probability density function (pdf) for λ_T is mapped through this curve to produce the corresponding pdf for power. Consequently, a shallow gradient produces a power pdf that is less spread out than a steeper gradient. At the optimum, by definition, the gradient is zero and there exists only one value that the expected power may take. The power from low friction channels (λ_0 is small) exhibits a greater sensitivity to the volatility in turbine drag as changes is turbine drag have a greater effect on the channel's dynamic balance. As the deployment of turbines increases and $\mu_{\lambda_T}/\lambda_0 \to \infty$, the ratio for the channel models GC05 and V10 approaches $1/2$ (uncertainty is reduced) while for GC13 it approaches unity (unit transfer of uncertainty).

Optimal Turbine Drag Another effect of uncertainty is that the maximum of expected power occurs at increasingly higher values of mean turbine drag upon introduction of uncertainty. This may be seen in Fig. 5.1 where the peak power shifts with increasing σ_{λ_T}. The optimal value of turbine drag is altered to maximise the

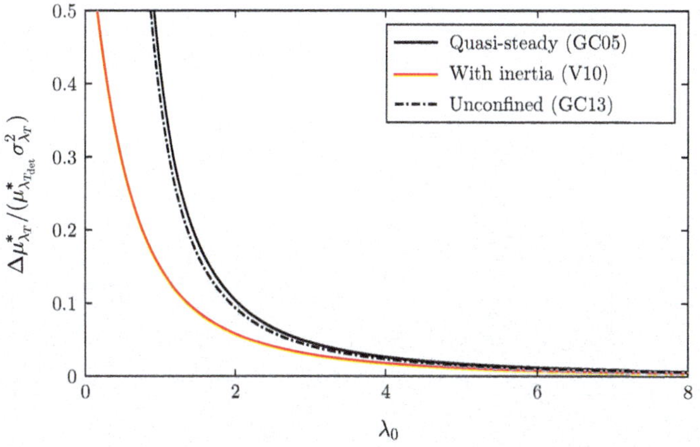

Fig. 5.5 Fractional change in optimal turbine drag scaled by turbine drag volatility as a function of bed roughness parameter

expected power out. To first order, the optimal mean turbine drag is

$$\mu_{\lambda_T,\mathrm{GC05}}{}^* = 2\lambda_0 + \frac{5}{6}\frac{\sigma_{\lambda_T}{}^2}{\lambda_0} + \mathcal{O}(\mathrm{E}[\Delta\lambda_T{}^3]). \tag{5.8}$$

Equivalently, for GC13, the optimal turbine drag is given by

$$\mu^*_{\lambda_T,\mathrm{GC13}} = 2\lambda_0 + \frac{3}{4}\frac{\sigma^2_{\lambda_T}}{\lambda_0} + \mathcal{O}(\mathrm{E}[\Delta\lambda_T{}^3]). \tag{5.9}$$

The relative change in turbine drag, along with that of V10, has been plotted in Fig. 5.5.

A general trend in the change of optimum turbine drag may be found by considering the stochastic optimum as a perturbation from the deterministic one, that is $\mu^*_{\lambda_T} = \mu^*_{\lambda_{T_{\rm det}}} + \Delta\mu^*_{\lambda_T}$. The perturbation may be found to leading-order by expanding the expression defining the stochastic optimum, $\partial\mathrm{E}[P]/\partial\mu_{\lambda_T} = 0$ about the deterministic value $\mu^*_{\lambda_{T_{\rm det}}}$ in order of the small perturbation. After some manipulation, this yields

$$\Delta\mu^*_{\lambda_T} = -\frac{1}{2}\frac{P_{\lambda_T,\lambda_T,\lambda_T}(\lambda_0,\mu^*_{\lambda_{T_{\rm det}}})}{P_{\lambda_T,\lambda_T}(\lambda_0,\mu^*_{\lambda_{T_{\rm det}}})}\sigma^2_{\lambda_T}, \tag{5.10}$$

where subscripts denote differentiation. As it is evaluated at the maximum, P_{λ_T,λ_T} is negative. More generally, the second derivative is in effect the responsiveness of power to uncertainty in λ_T and is plotted in Fig. 5.6 for the three models with respect to turbine drag. From the figure, it is clear that the third derivative $P_{\lambda_T,\lambda_T,\lambda_T}$ is positive

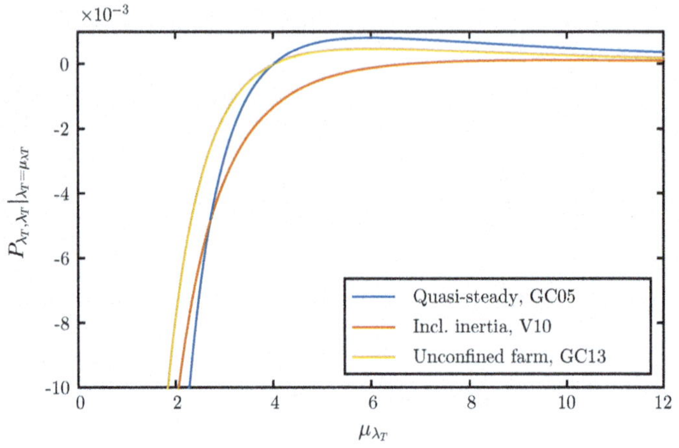

Fig. 5.6 Responsiveness of power to turbine drag uncertainty measured as second derivative of power with respect to turbine drag as a function of mean turbine drag uncertainty for a channel with $\lambda_0 = 1.0$. For the GC05 and GC13 models, the optimum (deterministic) turbine drag is at $\mu_{\lambda_T} = 2\lambda_0 = 2.0$

near the deterministic optimum. From (5.10), the change in optimal turbine drag is thus positive.

An under-designed channel, one with turbine deployment less than the optimum, is extremely sensitive to changes in μ_{λ_T}. The magnitude of the responsiveness reduces as more turbines are added. A channel with a greater deployment of turbines exhibits a smaller response to turbine drag uncertainty. Hence the optimal deployment of turbines, when the actual value of turbine drag is uncertain, is greater than predicted by deterministic models as the responsiveness of the power to volatility in turbine drag reduces. For very large values of μ_{λ_T}, the gradient in Fig. 5.6 becomes negative, however this is at turbine deployments far beyond the optimum and likely to be unfeasible.

5.2 Quantitative Estimates

The effect of uncertain turbine drag is next determined in terms of the change in power as a percentage of deterministic power and the percentage volatility in the power. To do so, an estimate for the relative standard deviation $\hat{\sigma}_{\lambda_T}$ is required. In all models, λ_T is a linear multiple of the turbine drag coefficient C_t. As a result, because the mean and the standard deviation of a random variable scale the same way if this variable is multiplied, $\sigma_{\lambda_T}/\mu_{\lambda_T} = \sigma_{C_t}/\mu_{C_t}$. To a first approximation, the uncertainty in turbine drag parameterisation is taken to be equivalent to that in bed roughness. Following the argument in Chap. 4, this is thus given by $\hat{\sigma}_{\lambda_0} = 0.41$.

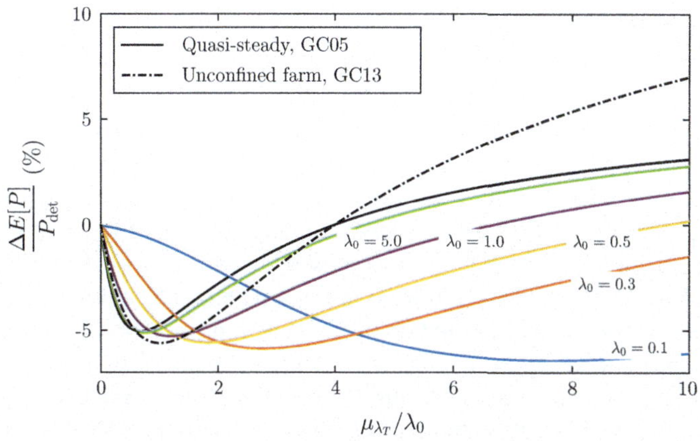

Fig. 5.7 Percentage change in expected power as a function of the mean turbine drag scaled by bed roughness parameter resulting from a turbine drag uncertainty of $\hat{\sigma}_{\lambda_T} = 0.41$ and for different channel dynamics in the V10 model

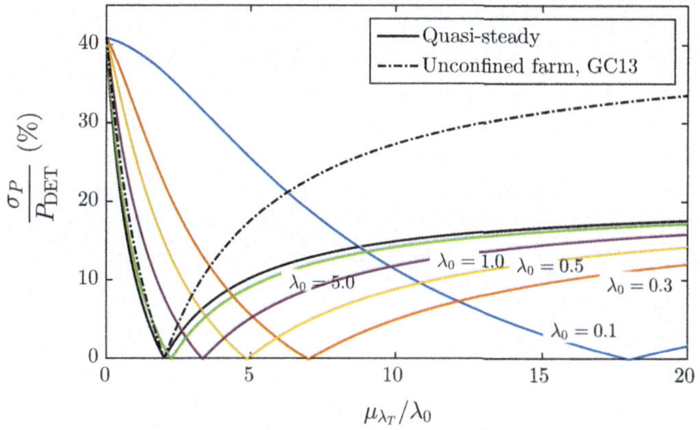

Fig. 5.8 Relative standard deviation in power due to turbine drag uncertainty ($\hat{\sigma}_{\lambda_T} = 0.41$) as a function of scaled mean turbine drag at different values of λ_0 for V10

Expected Power Figure 5.7 shows the change in expected power as a percentage of the deterministic power for a relative turbine drag uncertainty of $\hat{\sigma}_{\lambda_T} = 0.41$. The largest reduction occurs for inertia-dominated channels ($\lambda_0 = 01.$) and amounts to approximately -6% of the deterministic power. Increases of up to 10% occur in the unconfined farm model. However, these occur at infeasibly large turbine deployments $\mu_{\lambda_T} \gg \lambda_0$. For small values of $\mu_{\lambda_T}/\lambda_0$, the reduction in expected power is on the order of 5%.

Standard Deviation Figure 5.8 shows the standard deviation in power as a percentage of deterministic power for a relative turbine drag uncertainty of $\hat{\sigma}_{\lambda_T} = 0.41$. For all models, the volatility in power is at a maximum at the very initial stages of turbine deployment where the relative standard deviation is as high as 40%.

5.3 Conclusions

Uncertainty from several sources may affect the power estimates for a potential site. This chapter has examined the effect of uncertainty in the correct representation of turbines via an enhanced bed roughness parameterisation. Three analytic solutions previously established in the literature and outlined in Chap. 4 were considered, namely [3] (GC05) who derive an analytic solution for the power removed by turbines fully spanning a channel in the drag-dominated limit, [6] (V10) who relaxes this limit by retaining the inertial term in the governing equation and deriving an analytic solution to an approximation of the governing equation by linearising the drag term, and [4] (GC13) who consider a laterally unconfined turbine farm which allows for bypass flow. By using perturbation methods on the solutions presented in these models, leading-order expressions have been obtained for the expected power, standard deviation in power, and optimal turbine drag (which maximises expected power). By applying these expressions to a quantitative estimate for turbine drag uncertainty, namely $\hat{\sigma}_{\lambda_T} = 0.41$, the following findings were made. First, for small-scale turbine deployment ($\mu_{\lambda_T} < \lambda_0$) turbine drag uncertainty will invariably act to reduce expected power and can do so by as much as -6% for small, friction dominated channels. Second, for significantly underdesigned channels ($\mu_{\lambda_T} \approx 0$) power standard deviation may be as high as 41%. This however drops off quickly as more turbines are added and the deterministic optimum is approached, where the standard deviation in power drops to zero. Finally, the optimal deployment of turbines is greater than predicted by deterministic models upon consideration of uncertainty. The value for the fractional increase in optimum turbine drag is a function of the variance of the turbine drag itself and of the value of channel bed roughness parameter.

References

1. Adcock TAA, Draper S, Houlsby GT, Borthwick AGL, Serhadlioğlu S (2013) The available power from tidal stream turbines in the Pentland Firth. Proc Royal Soc A 469:1471–2946
2. Adcock TAA, Draper S, Nishino T (2015) Tidal power generation - a review of hydrodynamic modelling. Proc Inst Mech Eng A 1–17
3. Garrett C, Cummins PF (2005) The power potential of tidal currents in channels. Proc Royal Soc A 461:2563–2572
4. Garrett C, Cummins PF (2013) Maximum power from a turbine farm in shallow water. J Fluid Mech 714:634–643
5. Kramer SC, Piggott MD (2016) A correction to the enhanced bottom drag parameterisation of tidal turbines. Renew Energy 92:385–396
6. Vennell R (2010) Tuning turbines in a tidal channel. J Fluid Mech 663:253–267
7. Vennell R, Funke SW, Draper S, Stevens C, Divett T (2015) Designing large arrays of tidal turbines: a synthesis and review. Renew Sustain Energy Rev 41:454–472

Chapter 6
Pentland Firth Study

6.1 Introduction

The Pentland Firth is located at the northern end of Great Britain, separating the Orkney Islands from the Scottish mainland (see Fig. 6.1). It conveys extremely fast tidal flows with velocities regularly exceeding 3 ms^{-1} making the Pentland Firth a prime candidate for tidal stream energy extraction [4]. Currents through the strait are driven by head differences between water levels in the Atlantic Ocean and the North Sea, resulting largely from the phase difference of the tidal elevation between the two bodies of water as the tide propagates around the Orkney Islands from one side to the other [2]. The flow is further accelerated by the constriction of flow passages due to the presence of landmasses.

Several assessments have been made of the power that may be extracted from the Pentland Firth (see Table 1.1 in Chap. 1). There exists significant discrepancy between the values obtained, largely due to uncertainty associated with model choice, i.e. model assumptions, model setup, power metric used, etc. However, some of the discrepancy may be attributed to uncertainty arising from other sources such as choice of parameter values and boundary conditions (see Chap. 1).

In this chapter, the numerical technique for probability distribution transfer outlined in Chap. 3 and the analytic expansion method derived in Chap. 4 are applied to a case study of the Pentland Firth, modelled using ADCIRC-CG, an open-source finite element solver of the shallow water equations [9–11]. The effectiveness of leading-order estimates of the expansion method is tested in terms of their ability to capture the effect of bed roughness uncertainty on power from an actual channel system with complex, coastal boundaries and bathymetry.

M. J. Kreitmair, *The Effect of Uncertainty on Tidal Stream Energy Resource Estimates*, Springer Theses, https://doi.org/10.1007/978-3-030-57658-5_6

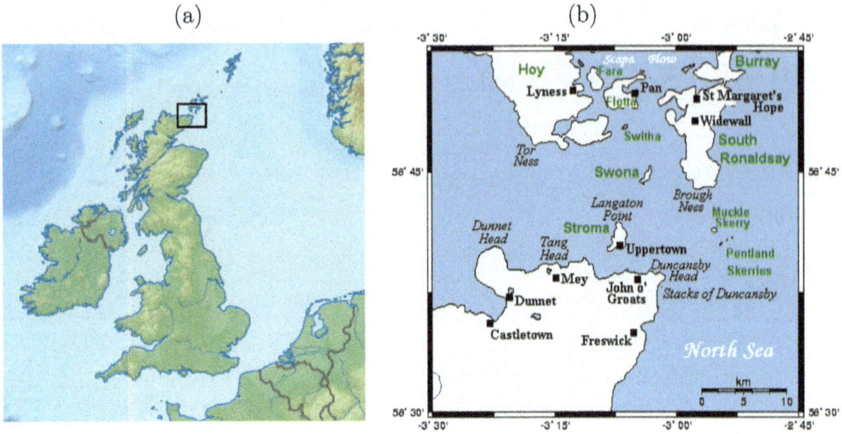

Fig. 6.1 **a** Location of the Pentland Firth within the United Kingdom and **b** close up. (*Source* Wikipedia commons)

6.2 Numerical Model

Numerical simulations for the Pentland Firth were performed using the Continuous Galerkin (CG) two-dimensional depth-integrated (2DDI) version of the ADvanced CIRCulation model (ADCIRC). ADCIRC is an open-source finite element solver which may be used to solve the shallow water equations, following [8], for application on boundary fitted, unstructured triangular grids.

The depth-averaged shallow water equations solved by 2DDI-ADCIRC may be summarised as follows. Mass conservation:

$$\frac{\partial \zeta}{\partial t} + \frac{\partial (hu)}{\partial x} + \frac{\partial (hv)}{\partial y} = 0, \tag{6.1}$$

where ζ is the free surface elevation, $h = h_s + \zeta$ is the total depth, h_s being the still water depth, (u, v) are the depth-averaged velocity components in the x and y-directions, and t is time. Momentum conservation (in conservative form), is given by

$$\frac{\partial (hu)}{\partial t} + \frac{\partial hu^2}{\partial x} + \frac{\partial (huv)}{\partial y} - fvh = -h\frac{\partial}{\partial x}\left[\frac{p}{\rho} + g(\zeta - \alpha\eta)\right] + \frac{\tau_{sx} - \tau_{bx}}{\rho} + M_x + D_x + B_x, \tag{6.2}$$

$$\frac{\partial (hv)}{\partial t} + \frac{\partial (huv)}{\partial x} + \frac{\partial hv^2}{\partial y} + fuh = -h\frac{\partial}{\partial y}\left[\frac{p}{\rho} + g(\zeta - \alpha\eta)\right] + \frac{\tau_{sy} - \tau_{by}}{\rho} + M_y + D_y + B_y, \tag{6.3}$$

where f is the Coriolis parameter, with $f = 2\Omega_E \sin(\phi) = 1 \times 10^{-4}\,\mathrm{s}^{-1}$ for the Pentland Firth ($\Omega_E = 7.29212 \times 10^{-5}$ rad s^{-1} is the angular speed of the Earth, and ϕ the degrees latitude), p is the free surface pressure, $g = 9.80665$ m s^{-2} is acceleration due to gravity, $\alpha = 0.69$ is the Earth elasticity factor, η is the Newtonian equilibrium tide potential, ρ is the density of water, $M_{x,y}$ the depth-integrated momentum diffusion, $D_{x,y}$ the depth-integrated momentum dispersion, $B_{x,y}$ the depth-integrated baroclinic forcings, $\tau_{sx,y}$ is the applied free surface stresses, and $\tau_{bx,y}$ the stresses at the sea bed (see [11] for more details). The bed stresses are assumed to be quadratic in flow speed and are be expressed by means of the (uncertain) bed roughness coefficient C_d via $\tau_b = (1/2)C_d u\sqrt{u^2 + v^2}$. In the derivation of these equations it has been assumed that the flow is essentially horizontal (i.e. uniform through depth, which is reasonable for a turbulent boundary layer), vertical acceleration can be ignored, the pressure distribution is hydrostatic, the seabed slopes are relatively small, turbulent eddies across different scales can be modelled using a single eddy viscosity coefficient, and tides may be modelled as long waves.

Figure 6.2a, b show the unstructured, triangular mesh that was fitted to the Pentland Firth. The mesh extends westward as far as the edge of the continental shelf in order to limit tidal reflections at the driving boundary [1]. The mesh was originally produced by Adcock et al. [2], after extensive convergence testing, with the size of the elements ranging from 300 m at the location of the turbine fences to 20 km far from the strait. The mesh resolution was chosen in a compromise between computational performance and accuracy.

The model was forced at the open boundary by prescribing time-varying water levels with an M_2 tide of angular frequency $\omega = 1.4052 \times 10^{-4}$ s^{-1}. While this

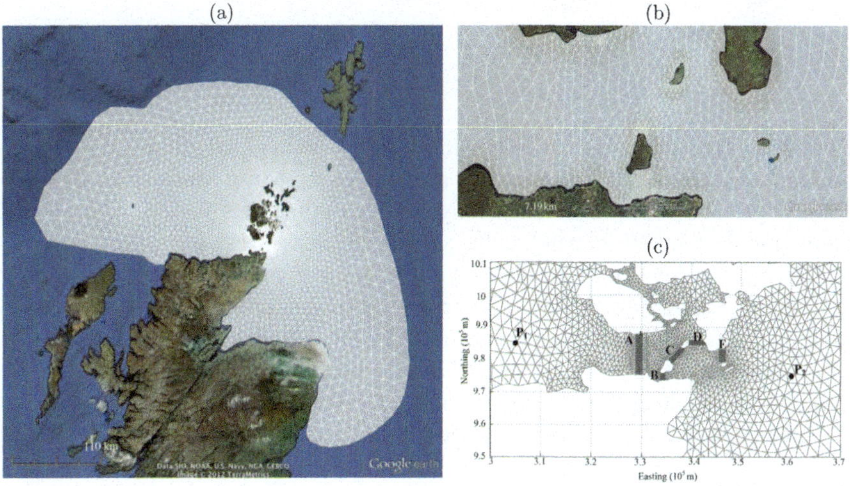

Fig. 6.2 **a** Mesh used in numerical solver, and **b** close-up of mesh at location of interest in the Pentland Firth overlaid onto a Google Earth image of the site (both taken from [2]); **c** areas of enhanced roughness designating turbine fences (taken from [4])

approach may give inaccurate results due to the incorrect calculation of mass flux at the open boundary [4, 7] showed that little change to the natural currents at the boundary occurs upon introduction of power extraction indicating that the 'clamped' boundary condition is appropriate. The model was run for two days, with a ramp-up 0.1 days before the average power over an M_2 tidal cycle was calculated.

Turbine Fence Representation and Configurations

Energy extraction by the turbines is modelled numerically using a local increase in bed roughness coefficient, equal to C_t, over areas representing the turbine fences. The areas considered are shown as strips labelled A-E in Fig. 6.2c. The increase in bed roughness is uniform over these areas, with each strip being approximately 1.5 km in width, as the power potential was shown to be stable for widths near this value [4]. Three fence configurations are considered in this chapter: A 'fully-spanned channel' configuration that deploys fences BCD, an 'asymmetrically confined' configuration that deploys fence B, and a 'laterally unconfined' scenario that solely deploys fence C.

The average power removed from the flow by the turbine fences over a tidal cycle is found by calculating the dissipation due to the increase in roughness over the turbine fence area, i.e.

$$\overline{P} = \frac{1}{T}\int_0^T \left(\int\int_{A_{\text{turb}}} \rho C_t \, |\mathbf{u}|^3 \, \mathrm{d}A \right) \mathrm{d}t, \tag{6.4}$$

where ρ is water density (taken to be 1027 kg m^{-3}), $T = 2\pi/\omega$ is the tidal period, A_{turb} is the plan area of enhanced bed roughness, and $\mathbf{u} = (u, v)^T$ is the depth-averaged velocity vector.

6.3 Results

The numerical model was run systematically for a range of values of bed roughness and turbine drag coefficients to create a matrix of the power removed by deployed fences for each of the different configurations outlined above. To obtain a finer resolution for the PDF transfer, cubic splines were applied to the power values for C_d and C_t, and the results interpolated onto a fine grid of linearly spaced values.

Figure 6.3 shows a selection of the resulting power surfaces (left hand subplots) and contour plots (right hand subplots) after application of the splines. The locations of optimal turbine drag (i.e. value that maximises the power dissipated for a given value of bed roughness coefficient), are indicated as a red curve. From these curves, it is evident that channel configuration and geometry affects optimal turbine drag, with

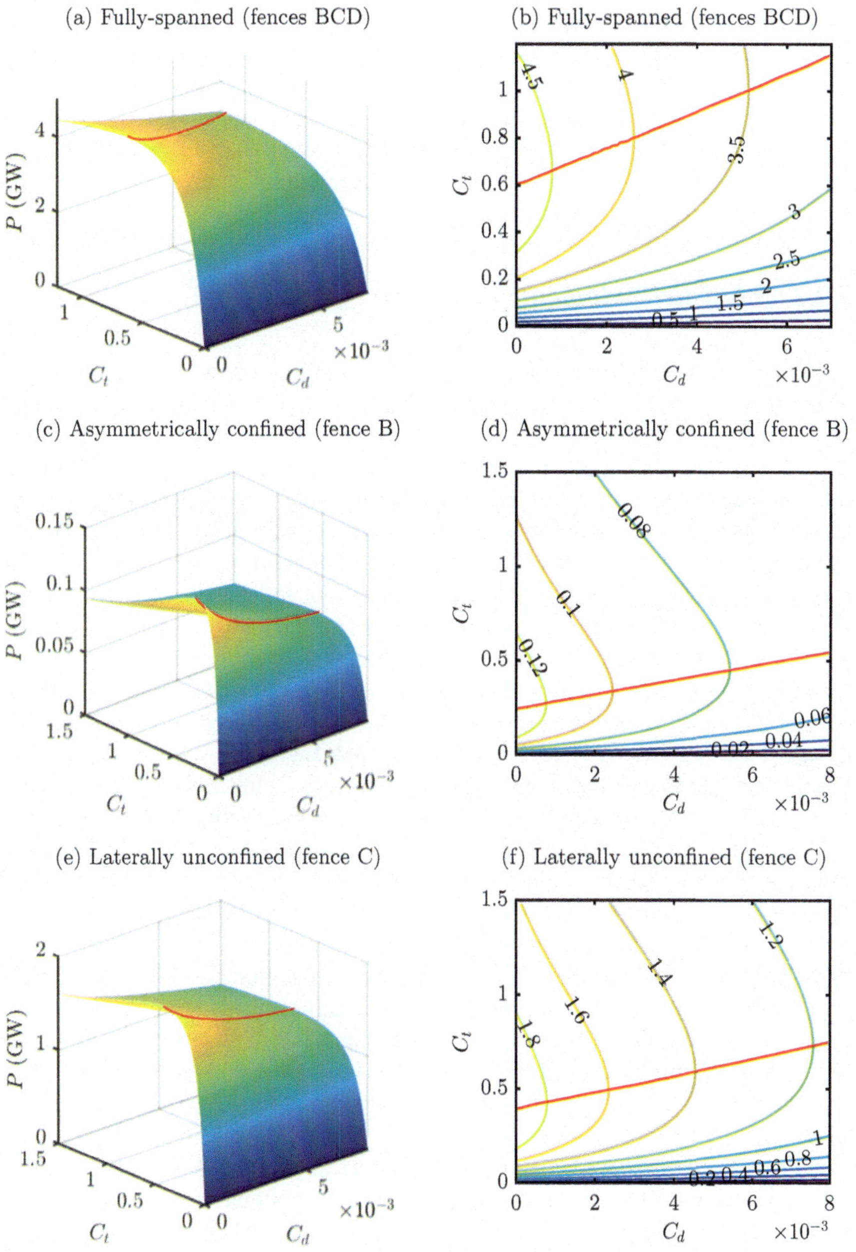

Fig. 6.3 Power extracted by fences BCD (panels **a** and **b**), fence B (panels **c** and **d**), and fence C (panels **e** and **f**) as a function of bed roughness coefficient C_d and turbine drag coefficient C_t. The red line indicates the location of maximum power dissipation

the largest value being exhibited by the fully-spanned fence configuration. The plots show that the power peaks for sub-channel deployments are steeper and exhibit larger gradients with respect to C_t (Fig. 6.3c, e) than power from a fully-spanned channel (Fig. 6.3a). This in turn affects how the expected power changes with uncertainty in C_d, because this change is, to leading order, dependent on gradients in the power surface (see Chap. 4), as is explored in the next sections.

6.3.1 *Statistical Moments for Power*

To determine the expected power for a given probability density function (PDF) of bed roughness coefficients with mean μ_{C_d} and standard deviation σ_{C_d}, the methodology proposed in Chap. 3 was employed. In summary, the PDF in C_d was transferred numerically through the power surface to determine the corresponding distribution for power and hence calculate the statistical moments. As before, the input distribution used is a symmetrically truncated normal distribution, chosen to prevent negative values of C_d from being assigned a likelihood of occurrence. The distribution was generated for different scenarios of mean bed roughness coefficient, $\mu_{C_d} = \{0.0025, 0.005, 0.016\}$, termed low, medium, and high channel drag scenario, respectively. The standard deviation of the truncated distribution (not the untruncated 'parent' normal distribution) was set to $\hat{\sigma}_{C_d} = \sigma_{C_d}/\mu_{C_d} = 0.41$, as in Chap. 4. For each of the three channel drag scenarios, the values of expected power and standard deviation for a given value of C_t were calculated. Figure 6.4 shows the results (as black curves) in terms of normalised power statistics as a function of the ratio of turbine drag to mean bed roughness coefficient. The left hand subplots show the change in expected power as a percentage of the deterministic power, the right hand subplots display the relative standard deviation in power. Each row of subplots presents the results for a different value of mean bed roughness coefficient. The locations of deterministic optima are indicated by open circles.

As may be seen from the figures, all fence configurations exhibit an increase in expected power, particularly at low values of relative turbine drag. As the turbine drag increases, the effect of uncertainty in bed roughness coefficient becomes smaller. This is because the power dissipated becomes less sensitive to the value of C_d, as indicated by the flattening of the power surfaces for all fence configurations shown in Fig. 6.3.

Of the three fence deployments, the asymmetrically confined fence configuration (black dashed curves) is most heavily impacted by uncertainty in C_d, indicated by a larger relative change in expected power compared to the other fence configurations for the low and medium channel drag scenarios. Moreover, the relative standard deviation in power σ_P/P_{det} is greatest for this deployment.

The laterally unconfined fence configuration (black dot-dashed curves, fence C), appears to be most affected by the change in mean bed roughness coefficient, first increasing by a factor of four as μ_{C_d} doubles from the low (panel a) to the medium

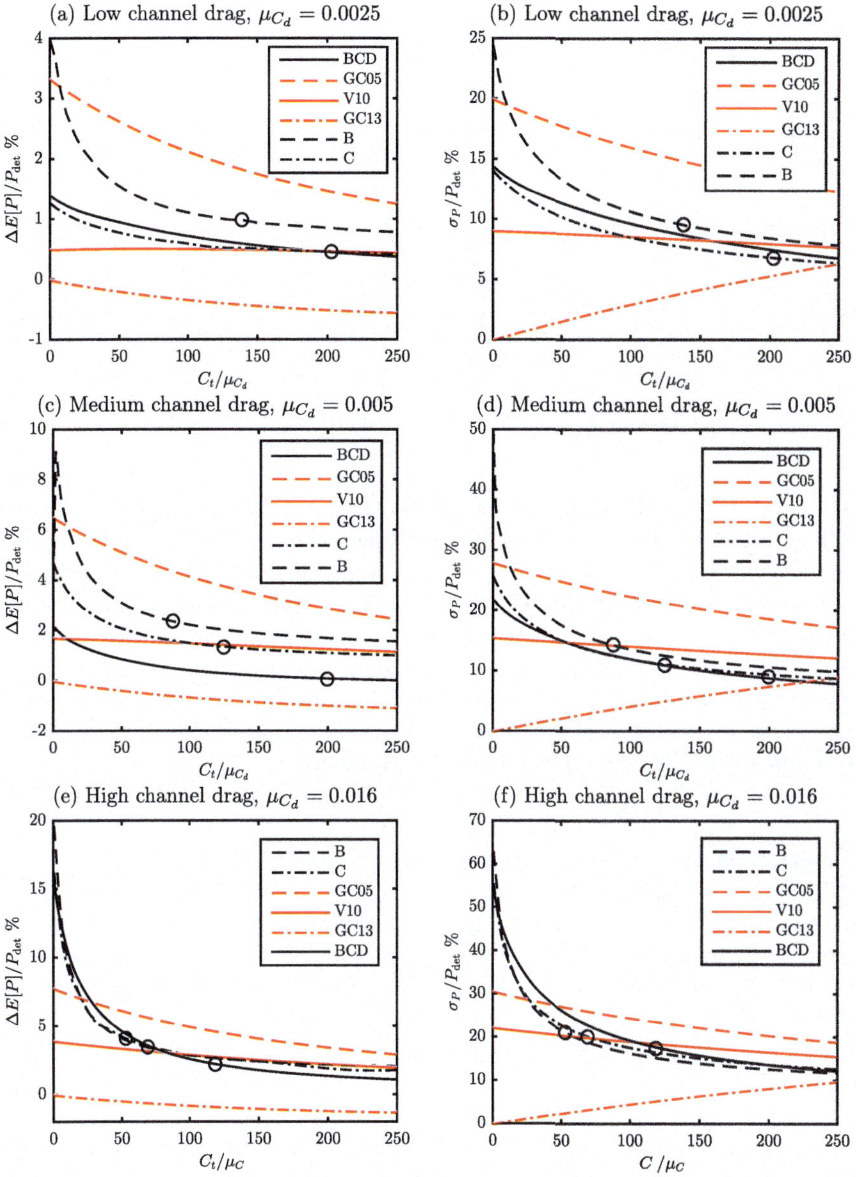

Fig. 6.4 Effect of bed roughness coefficient uncertainty on expected power (left column of panels) and standard deviation in power (right column of panels) for different turbine fence configurations (BCD is continuous, B is dashed, and C is dot-dashed black lines) and values of mean bed roughness coefficient. The locations of deterministic optima are indicated by open circles (location for BCD in panel a is off the plotted axis). Red lines designate results calculated from the expansion method for the models GC05 (dashed), V10 (continuous), and GC13 (dot-dashed plots)

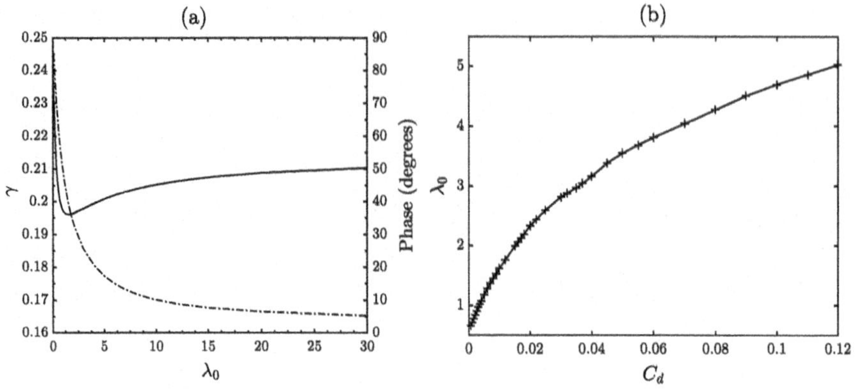

Fig. 6.5 **a** λ_0 as a function of phase difference in a channel (reproduced from GC05), and **b** λ_0 as a function of C_d

(panel c) channel drag scenario and then more than tripling as the mean bed roughness coefficient itself triples as we move to the high channel drag scenario (panel e).

At high channel drag, the plots for percentage change and for relative standard deviation for all three fence configurations approximately collapse onto each other. This indicates that the sensitivity of fence configurations, i.e. the gradient of power on C_d, becomes similar as we move to regions on the power surfaces at high values of C_d. An interpretation of this is that, high channel drag, despite fluctuations, acts to confine flow and helps funnel it into the turbine fences.

Comparison Against Simplified Analytic Models of Chap. 4

To test the ability of the analytic models to determine power from channels with complex bathymetric features and sub-channels, the analytic models of Chap. 4 are compared to the numerical results determined above. Overlaid onto the numerical results (black lines) in Fig. 6.4 are results calculated using the expansion method from Chap. 4 (red lines) for the different models of Garrett and Cummins [5] (GC05, quasi-steady flow), [12] (V10, approximate solution with inertia retained), and [6] (GC13, laterally unconfined flow). The values for μ_{λ_0} and $\lambda_T/\mu_{\lambda_0}$ associated with a given μ_{C_d} and C_t/μ_{C_d} were determined from the numerical model as follows.

As shown by GC05, the phase lag of the peak volume flux behind the dynamic driving head difference along the length of the channel uniquely defines the value of λ_0 for this channel, owing to the relative strength of inertia and bed roughness acting in the channel. Hence, for a known phase lag, λ_0 can be determined using Fig. 4 in GC05, which has been replicated in this thesis as Fig. 6.5a. The phase lag for the Pentland Firth for a given value of C_d determined by performing numerical simulations using the Pentland Firth model without turbines and analysing the time series for the head difference across the channel and the flow rate. The head difference

Table 6.1 Value of mean channel drag parameter μ_{λ_0} for a given mean bed roughness coefficient μ_{C_d} and a relative standard deviation of $\hat{\sigma}_{C_d} = \sigma_{C_d}/\mu_{C_d} = 0.41$

	Linear transfer		Non-linear transfer	
	μ_{λ_0}	σ_{λ_0}	μ_{λ_0}	σ_{λ_0}
Low channel drag, $\mu_{C_d} = 0.0025$	0.90	0.12	0.90	0.12
Medium channel drag, $\mu_{C_d} = 0.005$	1.18	0.22	1.17	0.21
High channel drag, $\mu_{C_d} = 0.016$	2.06	0.42	2.02	0.45

across the Pentland Firth was measured by subtracting the elevation time-series at point P1 (58°44′31.82″ N, −3°38′25.05″ E) from that at point P2 (58°39′39.53″ N, −2°41′28.21″ E), see Fig. 6.2c. These points are sufficiently far outside the channel that the head difference is no longer affected by either a jet exiting the channel or flow being 'sucked' into it. The flow rate was measured across a central transect of the channel. The time series was least-squares fitted with a sine series and the phase differences subtracted. Results are shown in Fig. 6.5b, which shows the dependence of λ_0 on C_d for the present Pentland Firth case study. This plot was used to determine the value of μ_{λ_0} for the three values of μ_{C_d} corresponding to the scenarios considered herein. Table 6.1 lists the mean and standard deviation of λ_0 for a given bed roughness coefficient scenario, determined assuming a linear relationship $\lambda_0 = \alpha C_d$ as in Chap. 4, and by transforming a normally distributed PDF for C_d using the graph of Fig. 6.5b. The values determined using the two methods are very similar to one another, due to the fact that the plot in Fig. 6.5b is locally well-approximated by a linear fit and the PDFs only sample a small portion of the graph at each $C_d = \mu_{C_d}$. As a result the values used for mean bed roughness parameter in the analytic results were $\mu_{\lambda_0} = \{0.9, 1.2, 2.1\}$.

Using Fig. 6.4, the leading order estimates from the expansion method may be compared with results determined from the numerical PDF transfer method. It is worth noting that the analytic models of GC05, V10, and GC13 are derived for simplified geometries, and that the expansion method yields only leading order estimates for the expected power and standard deviation. Nonetheless, the approach yields results of the same order of magnitude as the numerical results. In particular, the models of GC05 and V10, appear to bound the change in expected power determined numerically for the low channel drag scenario. V10 gives good agreement with most fence configurations for large values of relative turbine drag C_t/μ_{C_d} for low and high channel drag scenarios, but not for the medium channel drag case. GC13, on the other hand, does not give good estimates (predicting a decrease in expected power) as it models a circular patch of roughness in fully unbounded flow, which is a different geometry to that considered herein for the case of the Pentland Firth. Discrepancies are particularly pronounced at low C_t/μ_{C_d} values, and the qualitative behaviour of the expansion method results for V10 also differs to that from the numerical calculations. This indicates that the analytic models are less adequate at estimating changes in power extracted at low turbine drag.

6.4 Conclusions

This chapter has examined the effect of bed roughness uncertainty on power estimates from a numerical model of the Pentland Firth, U.K. in terms of change in expected power and standard deviation. These results were compared with leading order estimates made using an expansion based method applied to closed form solutions for power determined from three analytic models, GC05, V10, and GC13, as introduced in Chap. 4. The numerical simulations of the effect of bed roughness uncertainty show an increase in the expected power, particularly for low values of turbine drag, which is also exhibited by two of the three analytic models, GC05 and V10. The third model, GC13, however does not capture the effects due to uncertainty, as it was derived for a different flow geometry, indicating that the behaviour of sub-channel fence deployments is still dominantly that of a closed channel rather than that of a free standing turbine farm. Further work to explore the impact of bed roughness uncertainty in multiply connected channels is therefore required and it might be useful to incorporate uncertainty into the electrical analogy model of Draper et al. [3] which is able to take account of tidal fences within sub-channels. The sensitivity of power, both in terms of change in expected power and magnitude of standard deviation, increases with increasing mean bed roughness coefficient. Finally, while the expansion model is able to estimate the magnitude of relative change in expected power and relative standard deviation calculated by using the numerical PDF transfer, it is not able to capture the trends in behaviour with changing turbine drag.

References

1. Adcock TAA, Borthwick AGL, Houlsby GT (2011) The open boundary problem in tidal basin modelling with energy extraction. In: EWTEC, pp 1–7
2. Adcock TAA, Draper S, Houlsby GT, Borthwick AGL, Serhadlioğlu S (2013) The available power from tidal stream turbines in the Pentland Firth. Proc Royal Soc A 469:1471–2946
3. Draper S, Adcock TAA, Borthwick AGL, Houlsby GT (2013) An electrical analogy for the Pentland Firth tidal stream power resource. In: Proceedings of the royal society a: mathematical, physical and engineering sciences, 470
4. Draper S, Adcock TAA, Borthwick AGL, Houlsby GT (2014) Estimate of the tidal stream power resource of the Pentland Firth. Renew Energy 63:650–657
5. Garrett C, Cummins PF (2005) The power potential of tidal currents in channels. Proc Royal Soc A 461:2563–2572

6. Garrett C, Cummins PF (2013) Maximum power from a turbine farm in shallow water. J Fluid Mech 714:634–643
7. Garrett C, Greenberg D (1977) Predicting changes in tidal regime: the open boundary problem. J Phys Oceanogr 7:171–181
8. Hesthaven JS, Warburton T (2008) Nodal discontinuous galerkin methods - algorithms, analysis, and applications, vol 54. Springer International Publishing, New York
9. Kubatko EJ, Bunya S, Dawson C, Westerink JJ, Mirabito C (2009) A performance comparison of continuous and discontinuous finite element shallow water models. J Sci Comput 40(1):315–339
10. Kubatko EJ, Westerink JJ, Dawson C (2006) hp Discontinuous Galerkin methods for advection dominated problems in shallow water flow. Comput Methods Appl Mech Eng 196(1):437–451
11. Luettich RA, Westerink JJ, Scheffner NW (1992) ADCIRC: an advanced three-dimensional circulation model for shelves, coasts, and estuaries. Report 1. Theory and methodology of ADCIRC-2DDI and ADCIRC-3DL. Technical report technical report DRP-92-6, dredging research program, US army engineers waterways experiment station, Vicksburg, MS, USA
12. Vennell R (2010) Tuning turbines in a tidal channel. J Fluid Mech 663:253–267

Chapter 7
Conclusions and Recommendations

7.1 Preamble

In assessing the power available from rivers and tides, it is important to take into account uncertainty that can affect the accuracy of resource estimates. Such uncertainty can arise from inexact specification of physical parameters (such as bed roughness, flow geometry, turbine drag, tidal components, etc.), mathematical modelling parameters (such as model assumptions, approximations, boundary conditions), and numerical parameters (such as grid resolution, time step). The thesis has focused on the effect of uncertainty in bed roughness coefficient and turbine drag on stream power estimates. The effect was quantified in terms of the expectation value and the standard deviation in power from fences of turbines modelled as enhanced bed roughness. This chapter summarises the main conclusions of the thesis and suggests avenues for further research.

7.2 Methodology for Uncertainty Transfer

Three methods were devised to study the effect of bed friction uncertainty in river and tidal power assessment.

The first method involved the expansion of a functional expression for power as a Taylor series in terms of small fluctuations in bed roughness coefficient, resulting in leading order changes to the expected power in the deterministic state. This method is useful because only the statistical moments for the bed roughness are required to determine the expected power and standard deviation without exact knowledge of the PDF. In particular, analytic expressions were obtained for the power statistics, commencing from existing non-dimensional forms of the depth-integrated shallow water equations.

M. J. Kreitmair, *The Effect of Uncertainty on Tidal Stream Energy Resource Estimates*, Springer Theses, https://doi.org/10.1007/978-3-030-57658-5_7

The second method involved the numerical transfer of the probability density function (PDF) for the bed roughness coefficient through to a PDF for power. This method can be applied where the functional dependence of power on bed roughness is unknown. Potentially, this is a powerful technique that can be applied as a 'black box' to a wide range of uncertainty propagation problems in hydraulic engineering (e.g. tidal power assessment, urban flood risk, pollution of water bodies, coastal flooding, sediment transport and morphological change).

The third method comprised an analytic PDF transfer which required knowledge of both the input PDF and the functional dependence of power. For a given input PDF and power function, this is mathematically the most exact of the methods considered.

A study was undertaken of the effect of bed friction uncertainty on turbine power assessment in steady, one-dimensional flow, representative of a river. An analytic expression was derived for power extracted, and comparison undertaken between results from the three uncertainty propagation methods. Excellent agreement was achieved between the numerical and analytic PDF transfer methods. The degree of agreement improved as the numerical mesh was progressively refined. Results from the expansion method were also in satisfactory agreement with the predictions by the two PDF transfer methods. Here, the agreement improved with increasingly higher orders included in the underlying series.

7.3 Effect of Bed Roughness Uncertainty on Tidal Power Assessment

The effect of uncertainty in bed roughness coefficient was explored using the expansion method applied to the following three analytic models available in the literature for tidal stream resource assessment: (1) the quasi-steady state channel model of Garrett and Cummins [2] (GC05) where a fence of tidal turbines completely spans the flow; (2) a relaxation of the steady-state assumption of GC05 where the inertial term is retained in the governing equation and an approximate solution found by Vennell [4] (V10); and (3) the laterally unconfined model of Garrett and Cummins [3] (GC13) where a circular tidal turbine farm in free shallow flow is modelled as a circular patch of enhanced friction.

All three models exhibited an increase in standard deviation as well as a shift in mean (expected) power upon introduction of uncertainty. In both the channel models, GC05 and V10, the expected power increased, i.e. the average power determined from a distribution of bed roughness coefficients was greater than the power calculated using the average value of the bed roughness coefficients; this was due to the nonlinear dependence of power on bed roughness. However, the increase in expected power reduced with increasing inertia in the channel; this was because bottom friction had less influence on the flow velocity as inertia became more dominant in the dynamic balance for the channel. The unconfined model of GC13, on the other hand, exhibited a decrease in expected power because the dependence of flow velocity through the turbine farm on bed roughness was fundamentally different to that of the channel models. In the unconfined geometry of GC13, bed roughness did not act solely to decrease the flow speed, but also to funnel the flow through the farm by impeding the by-pass flow.

A case study was undertaken for the Pentland Firth using numerical data provided by an existing, validated depth-averaged shallow flow model (CG-ADCIRC) fitted to the ocean around the north of Scotland. Expected power values were calculated using the numerical PDF transfer method applied to the power surface (linking bed friction to power estimates) obtained using the numerical model. A comparison was made between these estimates and the expected power determined using the expansion method applied to the three analytic models. Close agreement in magnitude of expected power was achieved between the numerical model, and GC05 and V10 analytic model results. However, the analytic models were unable to predict the general trends in the relative change in expected power with varying turbine drag. Results from the GC13 model were not in good agreement with the numerical model, indicating that the behaviour in a multiply connected channel corresponds more to flow in a fully-spanned channel rather than a unconfined farm.

7.4 Effect of Turbine Drag Uncertainty

In addition to uncertainty in bed roughness coefficient, the thesis has also assessed the impact of uncertainty in correctly parameterising the turbine drag. Essentially the same expansion technique as used previously for bed friction uncertainty was applied, this time for turbine drag uncertainty, to the GC05, V10, and GC13 analytic models. By considering small perturbations in turbine drag, expressions for expected power and its standard deviation were derived. From the expressions, the following conclusions can be drawn.

- For small-scale turbine deployment (i.e. mean turbine drag smaller than bed friction), turbine drag uncertainty would reduce expected power and could do so by as much as -6% for small, friction dominated channels.
- For significantly under-designed channels (mean turbine drag approaching zero), the power standard deviation could reach 41%. This however would drop off rapidly as more turbines are added and the deterministic optimum is approached, and the standard deviation in power drops to zero.
- The theory predicts that the optimal deployment of turbines should be greater than predicted by deterministic models when turbine drag uncertainty is considered. The fractional increase in optimum turbine drag is a function of variance in turbine drag and the dynamic balance in the channel.

7.5 Recommendations for Further Research

The following bullet points provide a non-exhaustive list of ideas for further research that could be carried out as a result of the present work:

- The methods for uncertainty transfer outlined in this thesis could readily be applied to other free surface flow problems such as river flow routing, urban flooding, and debris flows, in each case accounting for uncertain terrain and terrain cover.
- The models for tidal power assessment considered in the thesis are highly idealised. More realistic models are worth investigation, such as linear momentum actuator disc models to represent the presence of tidal turbines. Such models allow for by-pass flow and can distinguish between extracted and available power, and could be used to track the effect of uncertainty on the local flow field and changes to the local environment (including sediment transport, bed morphological change, contaminant transport, water quality, and other ecological indices).
- The comparison study for power uncertainty in the Pentland Firth showed that analytic models do not accurately predict uncertain tidal stream power extraction from multiply connected channels. By considering a power model that includes one or more sub-channel(s), such as the electrical analogy model derived by Draper et al. [1], the effect of uncertainty at one tidal site on the power resource of another could be explored, which could have important impacts of tidal site leasing strategies.
- One feature which has not been considered is the possibility of spatially varying bed roughness coefficients and associated uncertainty. These could be incorporated, with knowledge of the coefficient decreasing with distance to a measuring point, by including a stochastic bed roughness coefficient within the governing momentum balance of the shallow water equations.
- Further research is recommended into developing analytic models that combine different sources of uncertainty (such as in bed roughness and turbine drag) using a multi-dimensional Taylor expansion. This would require knowledge of the correlations between the different random variables.
- The present work has considered forward propagation of uncertainty. It would be of considerable use for risk and reliability analysis, to carry out back-fitting exercises whereby the PDF of the 'input variable' (bed roughness coefficient, say) is determined from the PDF of the 'output variable' (power, say).
- As more data become available from satellite and field observation campaigns, it should be possible to obtain better estimates for PDFs of bed roughness coefficient in different rivers and coastal basins where stream turbines may be sited. This should also enable better fitting of input probability density functions (contenders being truncated normal, lognormal, etc.). This thesis has concentrated on the leading order statistical moments (i.e. expected value and variance). Future effort should be put into evaluating higher order moments (e.g. skewness, kurtosis, etc.) that shed light on the changes occurring in probability distributions as uncertainty propagates through the system being modelled.
- With the availability of increased computer power, more efficient computational algorithms, and large data computing (perhaps exploiting the cloud, GPUs, etc.),

it is recommended that engineers develop similar approaches to those used herein but for spatially three-dimensional flows past river and tidal turbines, perhaps using computational fluid dynamics, blade element methods, etc., to improve local resolution. Such models could generate petabytes of output, and so this would be an area ripe for application of data analytics techniques (especially methods that extract simple formulae to represent very complicated patterns of behaviour).

References

1. Draper S, Adcock TAA, Borthwick AGL, Houlsby GT (2013) An electrical analogy for the Pentland Firth tidal stream power resource. In: Proceedings of the royal society a: mathematical, physical and engineering sciences, 470
2. Garrett C, Cummins PF (2005) The power potential of tidal currents in channels. Proc Royal Soc A 461:2563–2572
3. Garrett C, Cummins PF (2013) Maximum power from a turbine farm in shallow water. J Fluid Mech 714:634–643
4. Vennell R (2010) Tuning turbines in a tidal channel. J Fluid Mech 663:253–267

The manufacturer's authorised representative in the EU is Springer Nature Customer Service Centre GmbH, Europaplatz 3, 69115 Heidelberg, Germany. If you have any concerns regarding our products, please contact ProductSafety@springernature.com

Printed and bound by CPI Group (UK) Ltd, Croydon, CR0 4YY

30/09/2026

02225471-0008